THE COMBAT SHOTGUN AND SUBMACHINE GUN

THE COMBAT SHOTGUN AND SUBMACHINE GUN

A SPECIAL WEAPONS ANALYSIS

CHUCK TAYLOR

Paladin Press
Boulder, Colorado

The Combat Shotgun and Submachine Gun:
  A Special Weapons Analysis
by Chuck Taylor

Copyright © 1985 by Chuck Taylor

ISBN (USA) 0-87364-312-7
Printed in the United States of America

Published by Paladin Press, a division of
Paladin Enterprises, Inc., P.O. Box 1307,
Boulder, Colorado 80306, USA.
(303) 443-7250

Direct inquiries and/or orders to the above address.

CONTENTS

ABOUT THE AUTHOR

To begin with, you must realize that this book is unashamedly honest, as it must be in order to make clear the points that may save your life in a gunfight. Taylor is cynical but not offensive while calling a spade a spade. The knowledge and experience that give him the right to be so have been earned and his credentials are impeccable.

In 1965 Taylor joined the Army and in that year he finished Advanced Infantry Training. By 1967 he had completed Infantry Officer Training School and Ranger training. He holds Military Expert Ratings with the .45 ACP M1911 pistol, 5.56mm M-16 rifle, 7.62mm M-14 rifle, .30 M-1 Carbine, 7.62 M-60 machine gun and 40mm M-79 grenade launcher. During 18 months' service in Vietnam, Taylor earned these decorations: Vietnam Service, Vietnam Campaign with five stars, the Bronze Star with "V" device, Vietnamese Cross of Gallantry with Palm, Combat Infantryman's Badge and the Purple Heart. He retired from the service with the rank of Captain, O-3, and still holds that commission. With these military qualifications as a basis, Taylor achieved civilian Expert ratings in NRA Light Rifle, High Power Carbine and High Power Rifle. He is an NRA Certified Rifle and Pistol Instructor, has held an IPSC Class A rating since 1976, and was a member of the 1978–1979 IPSC United States Team.

Jeff Cooper declared Chuck Taylor a *Master Marksman* early in 1979 when he joined the staff at the American Pistol Institute as Operations Manager and Senior Instructor. Taylor resigned from his post at Gunsite in the latter part of 1980, and shortly after founded the American Small Arms Academy. Operating under that banner, Taylor has become one of the most sought-after shooting instructors in the world, not only by civilians from all walks of life, but from elite organizations throughout the Free World. Africa and Latin America have felt the impact of his work as a small arms consultant, and members of many SWAT and TRU teams, U.S. Special Forces/Ranger units, the Rhodesian SAS, Israeli Army and German GSG-9 are alive today as a result of his training. Taylor became the only known *Four Weapon Master* (rifle, shotgun, submachine gun and handgun) in 1981.

It is clear that Taylor has elevated practical weaponcraft to a new level: that of a discipline for self-defense and personal protection, as are other martial arts and sciences that are applicable to the control of our environment. His teachings inspire pride and confidence in one's self and one's accomplishments with a firearm. The weapon is taught with an eye to the dignity owed a tool, an object of pleasure, and an instrument of life.

CHRIS McLOUGHLIN
Associate Editor
SWAT magazine

PREFACE

There is little question that the two weapon types discussed herein are greatly over-, and at the same time, underestimated in terms of their flexibility and general utility. No other two categories of small arms elicit such a strong emotional response from their advocates, sometimes bordering on physical confrontation. This disturbs me greatly, because there is almost no basis for comparison between them, no common denominator except that they are both firearms, tools in the combat sense.

All too often, heated debate erupts in print and word over which arm, the SGN or SMG, is "the best"; such endeavors can only be viewed as an exhibition of the ignorance of those involved. I use the term "ignorance" not with demeaning intent, but as a definitive expression. I must do so, for while an academic comparison of the two arms will always show the SMG to be far more versatile in combat applications than the shotgun, one could also safely state that a rifle is more effective than either one. Weapon effectiveness depends upon the mission for which the weapon will be employed, does it not?

Unfortunately, this criterion usually falls immediately by the wayside in such uninformed discussions. The temptation to compare our pet shotgun with that SMG we know little or nothing about is often irresistible, even for some so-called "experts." Legal restrictions on the SMG are far more prohibitive than on the shotgun, and this influences greatly its accessibility to interested personnel. Thus, comparisons between the two guns invariably revolve around assumed and/or illegitimate theoretical concepts with virtually no attention paid to historic precedent or practical reality.

The generality of "gun writers" fall into this category because few of them are actual professionals. As surprising as this might seem, a visit to their armories turns up little more—and often less—than you have in your own. The difference is that they have a typewriter and a camera.

This is not the case with me or my armory, which contains examples of every major SMG and SGN design in existence. Attendance at one of my T&E or practice sessions will confirm that I am not a hobbyist with a typewriter. Rather, I am a professional, one who for a living deals on a daily basis with weapons and tactics in the real world. This is why I so often disagree with those who either knowingly or unknowingly present to an unsuspecting public data which I know to be invalid. I feel a responsibility to my clients, readers and associates to ascertain the truth before picking up the pen. I could not satisfy my own code of conduct were I not to accomplish this, devoid of the incompetence and egotism I see daily in this business.

I make no claims to perfection, but I do claim perfection as my goal. To strive for less would be both dishonest and irresponsible in a business that naturally demands perfection. After all, how much more serious than life and death can things get? An invalid concept or technique offered out of ignorance or dishonesty can quite literally cost you your life; the field of battle is not the place to discover that your pet expert didn't know what he was talking about.

The material that appears in this book is presented in the spirit of saving lives, however the cards may fall. It is compiled from years of personal experience, interview and historic analysis. Opinions are clearly labelled as such with supporting reasons.

In short, in this, my third and final book about combat weaponcraft, I offer the serious student/reader a perspective on a frustrating and often confusing subject, but one no less important than the rifle and handgun. As a mortal man, I can do no more, except keep on trying. If my concepts and techniques are wrong, then I, too, will die—but I haven't yet and neither have my students. This in itself is formidable testimony that I am at least on the right track.

An objective perusal of this book will confirm and disprove many things to you, some of which you have probably suspected and others which may surprise and even shock you a bit. However, as disturbing as this may be, remember that it is, above all else, the truth.

Remember too that in the final analysis, the individual, far more than his weapon, influences the outcome of a deadly encounter. If the submachine gun or combat shotgun are used for the purposes for which they are intended, with common sense, proper tactics and technique, they are formidable, effective arms. If not, they are virtually worthless.

All too often in our technology-oriented society we forget this absolute. Most who do don't live to tell about it. Too often even those who remember it die, as Capstick says, "in the silent places." I've been to those places—and there are many of them—and seen it firsthand. It is this firsthand experience that prompted my initial quest to "find a better way," because a great many of those who fell were my friends, people I grew up with, laughed and cried with, and drank Jack Daniels with; people of the mold that God threw away long ago—the old breed ... my breed.

CHUCK TAYLOR
Prescott, Arizona

FOREWORD

At the risk of disappointing some, I'll forego the requisite "Chuck Taylor needs no introduction . . ." and the usual follow-up listing of his qualifications. While our professional and personal association of many years has been both a pleasure and an education for me, I suspect that Chuck's reputation in the world of serious weaponcraft has far surpassed even his own realization. If you are taking the time to read this, you have most likely heard of the man, read his material in SWAT magazine or other periodicals, or experienced his tutelage as an American Small Arms Academy student.

Instead, I will give you a brief perspective on something which does need introduction—desperately so in the case of many uninformed or impressionable readers: the subject matter of this, Chuck's third book.

I think it safe to say that no two weapons of modern times are more thoroughly misunderstood and misemployed (by people who ought to know better) than the fighting shotgun and the submachine gun. Neither firearm is new on the combat scene and, moreover, neither is going to go away if we close our eyes and wish it so. Carried by semi-literate peasants and Secret Service men alike, by the elite of the world's armies and by the best (and worst) of law enforcement agencies, these weapons are here to stay. Neither one is a "slob's weapon" (if such a thing does exist, it must surely be the typewriter!), and yet, neither is omnipotent—champion skeet shooters and Hollywood screenwriters notwithstanding.

Rather, both are tactical tools; lethal tools to be sure, in the hands of trained men but, nonetheless, two of many means whereby the combative problem may be solved and one's enemy neutralized. The fact that they seem the best such tools in a number of scenarios is both easy to deny for reasons of personal prejudice and easy to prove, once one has attained a respectable level of skill. I have witnessed firsthand on a score of occasions what Chuck Taylor can do with either arm. More important, I have come to understand the how and why of his ability; and that is the chief purpose of this final volume in his weapons trilogy: to put this same deadly skill within reach of the responsible and interested reader. You will find the techniques of the combat shotgun and the submachine gun neither extremely difficult nor extremely easy to master. But any serious reader will find both weapons eminently effective when properly employed in their fighting roles.

If, as do so many firearms enthusiasts, you must be lavishly entertained by what you read, I suggest you look to another volume featuring dramatic photographs of bursting fruit and exploding water jugs. Better yet, use your hard-earned money to take in a good war movie or two.

But if it's education and instruction you're after, a book that can give you the "right stuff" on a controversial topic, look no further.

You hold it in your hands.

ERIC STRAHL
Associate Editor
SWAT magazine

1. THE RIGHT TOOL FOR THE JOB

While regarded by some as a mundane subject, it is necessary to examine this portion of both the submachine gun's and shotgun's personality because both are the result of prolonged evolution. And, although they share certain theoretical applications, they are as different as night and day in a far broader sense. In order to fully understand them, we must look into the past.

The shotgun is one of the simplest of firearms. It is a smooth-bored shoulder arm, launching a cargo of multiple projectiles. In fact, the history and development of the shotgun generally parallels that of rifled arms until the appearance of the repeating rifle in the 1850s. At this point, the SGN became essentially a provider of fowl and small game, and remained so to modern times, even though the advent of smokeless powder around the turn of the century opened new vistas of weapon development.

A number of important advances in firearms technology took place during this period, the self-loading handgun being one example. But without question, the most significant development was the machine gun, complete with belts, tripod and cumbersome water cans. While hardly portable by present-day standards, the MG did possess a singularly important capability—sustained, powerful long-range fire. The MG revolutionized warfare overnight and a long, bloody period of tactical innovation was required to fully understand its role in modern combat.

While all of this was taking the ordnance world by storm, the shotgun was quietly relegated to a dusty corner of the weapons closet, with little practical development beyond metallurgy and manufacturing methods.

From an objective standpoint, it is easy to see why this happened. The concept of the shotgun is so simple that it was deemed suitable only for minor, short duration scenarios, and then only as a weapon of necessity. True, the SGN did see some action in Pershing's Mexican incursions and in the trenches of World War I. But outside of highly specialized operations, it remained largely a civilian arm used in some law enforcement applications and a great many sporting endeavors.

In fact, that the shotgun was used at all during World War I calls graphic attention to the recognition by some military planners that modern warfare had dealt them some jokers as far as weapons and tactics were concerned. Specifically, it became obvious that once the rush across No Man's Land was accomplished and action shifted to the close confines of trenches and fortifications, often during nighttime hours, the conventional "tools of the trade," the rifle and pistol, were somewhat inadequate to deal with numbers of armed adversaries at close quarters.

Familiarity with the shotgun's basic concept caused it to be resurrected and shoved into the arena as a "trench gun." In this role it served relatively well, although its inherent fragility under military conditions, heavy recoil and excessively bulky ammunition, quickly made its limitations apparent. Nevertheless, the shotgun was popular in trench raids and other close actions if for no other reason than it was the only tool available that even marginally fulfilled a pressing tactical need. Today, more than six decades later, the SGN suffers from these same faults, which is why its military employment has remained minor while other weapon concepts have blossomed.

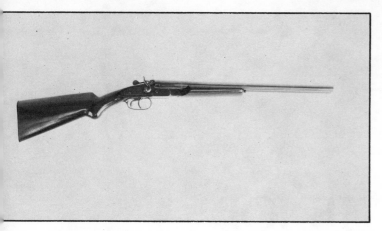

First generation SGN, side-by-side, double-barrelled, break-open. Pictured is the Rossi Overland.

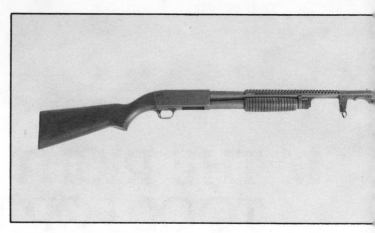

Second generation SGN, slide action. Illustrated here is the Ithaca M-37.

More important than its limited military application is the shotgun's role in civilian home defense and law enforcement missions. Being one of our oldest firearms, it has seen tremendous public exposure and, because of this and a number of misconceptions about its power and versatility, it has gained a solid footing in residential and urban personal defense. There is some justification for this popularity, for in such a role, the shotgun's limitations are no liability and are, in some situations, actually an asset.

Here we speak specifically of the SGN's lack of range and the minimal penetration potential of shotshells. Too, the SGN's gaping muzzle has significant psychological impact upon anyone at whom it is pointed, born in part from its reputation for great power. The SGN is quite capable of inflicting incapacitating or lethal wounds at short ranges and, as previously mentioned, its low penetration capability is as much of an asset for home defense or law enforcement as it is a liability for military operations. While even handgun bullets can easily penetrate multiple walls and other obstacles, shot is easily stopped by many obstacles commonly found in urban areas. This characteristic is clearly a legal and tactical asset, and can be further enhanced through the use of lighter shot with almost no loss in short-range effectiveness. A further advantage of light shot loads is reduced muzzle flash and recoil, important factors in training and use of the weapon.

Other than the recoil and gas-operated self-loading designs that were the result of innovations in automatic and semiautomatic rifled arms, the only real advancement in shotgun technology came in the late 1800s with the advent of the slide-action. Preferred by many for its operational sim-

plicity, the slide-action ("pump") continues to ride a swell of popularity to this day and is generally regarded as the state of the art of shotgun development.

Of significant interest are the few attempts to "update" the SGN into a truly modern arm. Such weapons are typified by the High Standard M-10 and Franchi SPAS-12, both of which utilize "high tech" metal stampings, phenolic resins and aircraft aluminum in much the same way as the M-16 rifle. Unfortunately, these weapons exhibit a considerable loss of perspective on the part of their designers who seem to dwell upon the marvels of mechanical engineering at the cost of practicality and efficiency. Perhaps for the same reasons, neither gun met with much success against the traditional "second generation" shotgun, epitomized by the Remington M-870, Ithaca M-37, Mossberg M-500, High Standard M-81100, and the Winchester M-97, M-12, and M-1200.

Fourth generation SGNs are characterized by the use of phenolics, aluminum, and other "high-tech" innovations. These guns have met with little commercial success. Pic-

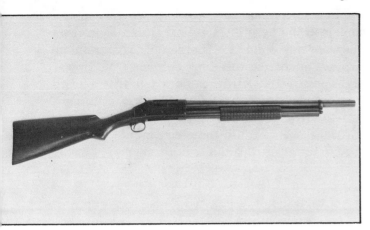

The Winchester M-97 is a second generation SGN.

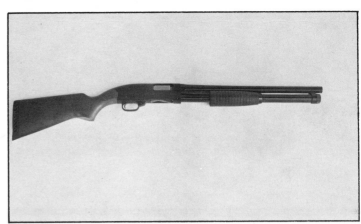

The Winchester M-1200 is another second generation SGN.

At the risk of offending those charged with selling "new generation" guns, I can shed some light on the matter with a simple observation—none of these "new" weapons offer any real increase in practical effectiveness over existing designs, most of which are far less expensive. This seems odd to me, for this should first be considered by anyone contemplating the design, manufacture, and marketing of any new product. However, attempts to create a job for the tool instead of a tool for the job run rampant in today's world.

Although technologically surpassed by the slide-action and self-loader, the "first generation" SGN, the side-by-side double and single barrelled break-open, remains fairly popular. This may be because they have been an American tradition for countless generations but more probably they continue to see service because they are readily available in economical "utility" versions and are simple to operate, especially under stress. This is,

of course, a great asset if the arm is to be used by semi- or untrained personnel.

Oddly enough, the third type of first-generation gun, the over/under double barrel, sees little combat use even though it is no less effective than any other SGN. The high cost of most of these is the likely culprit here. The O/U is a "gentleman's toy," most often seen on trap and skeet ranges and in the mahogany panelled gun rooms of the upper class.

Another design that has met with only marginal success is the self-loader, commonly (and inaccurately) referred to as an "automatic." Of the relatively few semiauto SGNs available today, the Remington M-1100 is by far the most prolific. Unfortunately, it is often unreliable when used with buckshot and slugs. This leaves only the now-defunct Browning A-5 and SKB XL series, as well as some Smith and Wesson models, from which to choose.

tured on the left is the High-Standard M-108, above is the Franchi SPAS-12.

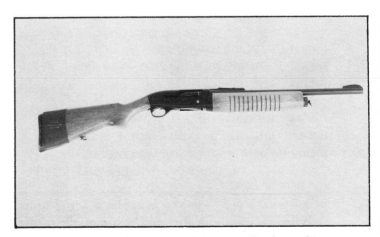

While far less popular than the slide-action because of its reduced reliability and scarcity, the auto-loading SGN is representative of third generation designs. Shown is the SKB XL-100, a reliable self-loader.

SMGs can be classified by generations. The first generation M-1928 Thompson, top, was the second practical SMG invented. The Soviet PPD-38, bottom, also reflects typical first generation characteristics.

I cannot comment on the relatively scarce Smith & Wesson, but I have seen enough unreliable performance by the Remington M-1100 to refrain from recommending it for serious use. I can also endorse the Browning A-5 for it is indeed reliable, if a bit hard on the shoulder because of its recoil operation. Also now in short supply, but highly reliable, is the SKB XL-100, a weapon which proves that gas-operated shotguns can be both economical *and* efficient.

At the same time the shotgun was being pressed into service during World War I, military planners were looking for a replacement for it. Reports of the shotgun's fragility and inability to handle problems requiring more range and penetration were common. In addition, complaints were received about the weight and bulk of its ammunition, which limited severely the quantities which its operator could reasonably carry.

It is interesting to note that two very important men in the ordnance field, Hugo Schmeisser and General John T. Thompson, were simultaneously pondering this same problem from opposing sides of the Atlantic. Both felt that a rifled arm, capable of fully automatic fire, utilizing a low-powered and thus easily controlled pistol cartridge, would fulfill the need for both short and moderate range capability. Both men were aware that such an arm, the Villar Perosa, had been fielded by the Italians in 1914 but it had suffered from being employed as an antiaircraft gun, not exactly a mission for which its 9mm cartridge was well suited! To further compound the confusion, the Italians mounted the weapon on the handlebars of a bicycle to increase its mobility!

Schmeisser was able to convince the German military hierarchy of the need for a moderate range, high volume of fire weapon and his creation, the MP-18, actually saw combat in 1918. Unfortunately, General Thompson was unable to generate much enthusiasm on this side of the Atlantic and was forced to seek private capital for the development of his concept, which he called a "submachine gun." Never a quitter, Thompson was able to perfect his SMG and quantities of the M-1919 version were actually stacked on the docks in New York City awaiting shipment to Europe when the war ended.

The period following World War I was a dismal one for arms development. The general opinion among the Allies was that since WWI was "The War to End All Wars," no further weapon research was needed. The Germans, on the other hand, hard pressed by the ridiculous Treaty of Versailles and already planning a comeback, were impressed enough by the performance of Schmeisser's MP-18 to secretly continue their SMG R & D program under the guise of developing a "machine pistol." Hence the title which has become a confusing generic term on the Continent—the machine pistol (MP).

Thompson, sadly, was cursed with a continued lack of interest in his weapon, which he refined into the M-1921 and M-1928 versions. In spite of limited use in 1928 by the Marine Corps Expeditionary Force to Nicaragua and extensive testing by the military that pronounced the gun to be excellent by all concerned, it remained largely ignored. The problem, the military said, was that the TSMG was neither a rifle nor a handgun, and thus was of no use! As typically occurs in this country after all wars, attention turned toward

Germans pioneered SMG tactics in World War I and II. Here an SS trooper of Colonel Otto Skorzeny's elite commandos practices glider dismount with his MP-40 SMG in preparation for the Mussolini rescue mission that stunned the allies.

winning past wars instead of future ones and innovation was stifled. It was not until a full decade later, in 1939, when the terrible folly of this attitude was to crash down upon us. By then, the Germans had already developed a second generation of SMGs and issued them in quantity to their troops.

First generation SMGs were characterized by so-called "Old World" craftsmanship and manufacturing methods which were both complex and time-consuming. From the outset, it was apparent that these traits were of little practical value for the SMG's intended mission and excessively expensive to boot. However, the technology of the day required such methods, for there was simply no other way available at the time. Hard at work to solve this problem, the Germans had by the late 1930s modernized their manufacturing techniques through the use of phenolic resins, spot welding, and stampings to the point where they were able to introduce the first example of the "second generation" SMG, the MP-38. Reflecting the desire for more compactness and lighter weight, the MP-38 saw considerable action in the Spanish Civil War and the early days of World War II and performed well enough to be highly regarded by those who used it. Minor improvements in its design and manufacturing procedures were made and its successor, the MP-40 or "Schmeisser" as it was to become erroneously known to Americans, became an overnight classic, much to the chagrin of Allied

ordnance officials and those who had to face it in combat.

By 1939 the United States was painfully aware of its inevitable involvement in World War II and its complete lack of a suitable SMG. In an attempt to field a stopgap weapon, the M-1928 Thompson, even then an antique, if a good one, was dusted off and retested. It natually passed its trials and was immediately standardized. But the TSMG, being a first-generation gun, suffered from all of the drawbacks of its breed—excessive expense, overly long manufacturing time and excessive weight—and the U.S. Ordnance Department immediately began a long-term search to find a replacement for it. This led to the M-3 "Grease Gun" which, perhaps because it was born in the shadow of the famous Thompson, never gained the reputation or popularity of its predecessor.

However, it is safe to say that secret evaluation of captured German MP-38's and MP-40's caused quite a stir at Aberdeen and the reaction they

Shown here is a German trooper with his MP-40.

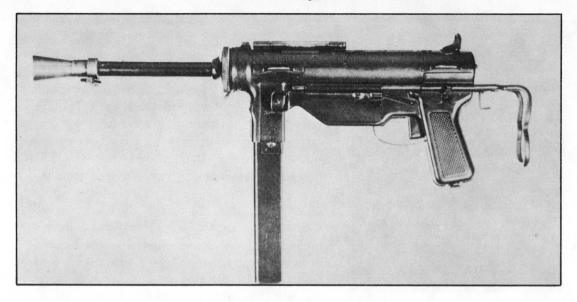

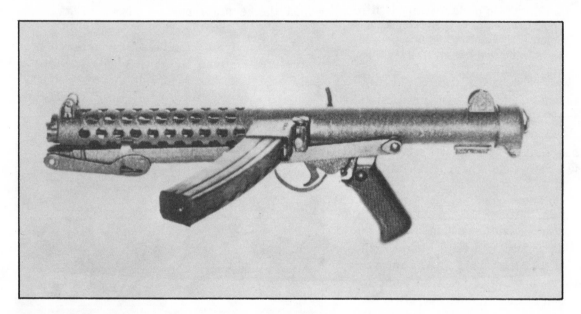

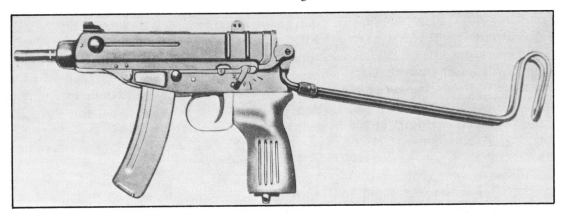

Second generation SMGs are typified by the use of extensive stamping and spot welds, as well as folding stocks to reduce bulk. Shown here are the U.S. M-3 "Grease Gun" (top, opposite page); the U.S. Sterling L-2A3 (middle, opposite page); the U.S. Madsen M-76 (bottom, opposite page); and from the 1960s, the Czech M-61 (top, this page); and the U.S. Smith & Wesson M-76 (above).

caused, bordering on panic, sparked great improvements in Allied SMG—and other small arms—policies. The British, in an even more difficult predicament after their stunning defeat at Dunkirk, were now quite ready to listen to reason and suddenly couldn't get enough SMGs, even though they had previously turned up their collective noses at the Thompson, dubbing it a "gangster gun" in reference to its colorful Depression-era past. They were to go on to make good use of their TSMGs in the years following the Battle of Britain; the Thompson even became part of the official insignia of the famed Commandos.

Based upon either U.S. or their own evaluations of the German MP-40, other nations developed their own second generation SMG designs. The STEN, Gustav M-45, Madsen M-50, MAT-49, and the later Smith & Wesson M-76 were all products of this effort.

The final days of World War II disclosed that the Soviets had used staggering quantities of PPSh-41 and PPS-43 SMGs with great success. They had learned about SMG effectiveness the hard way

during the Russo-Finnish War, when entire units of Russian infantry were wiped out by Suomi SMG-toting Finnish ski troops. Swooping down upon them from the heavy, snow-laden timber of the Finnish countryside, the Finns inflicted horrendous losses on the Russians, who won the campaign after paying a terrible price in lives.

Toward the end of the war, the Germans had perfected another new concept in small arms, the Sturmgewehr, which was touted as the successor to both the conventional battle rifle and the SMG, and used with telling effect against the Soviets at Stalingrad. The Wehrmacht began rearming its troops with the new rifle "too little, too late" to have much effect as advancing Russian troops quickly overran Eastern Europe. With peace (sort of) at hand, ordnance experts the world over began taking stock of the events of the preceding five years and realized the theoretical potential of the "assault rifle," which utilized an intermediate-powered cartridge and thus fulfilled the role of both the rifle and SMG. As a result, simultaneous efforts to produce new versions of the weapon

The Thompson SMG was so popular among British Commandos that they actually refused to trade in their M-1928 TSMGs for Mk II STENs! Here three Commandos interrogate two captured German prisoners. Two of the three are armed with Thompsons.

An added advantage of the SMG is its ability to perform special functions. Silencers and sound-suppressors add considerably to this capability.

The Ingram-10 with suppressor.

The MK IIS STEN with suppressor.

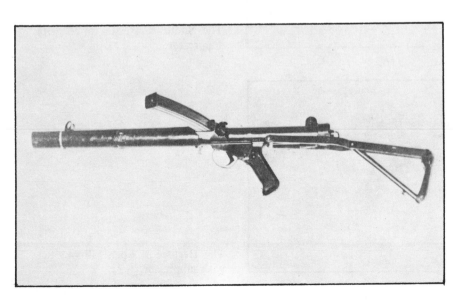

The Sterling L-34 with suppressor.

The Gustav M-45 with suppressor.

The M-3 "Grease Gun" with suppressor.

The Smith & Wesson M-76 with suppressor.

The Heckler & Koch MP-5A3 SD with suppressor.

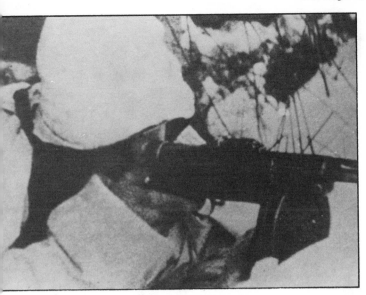

A Finnish soldier in winter camouflage fires his Suomi at Soviet troops during the Russo-Finnish War of 1940. The Finns actually caused mass panic among Russian infantry with their hard-hitting attacks utilizing mass SMG tactics.

Soviet troops in action with their PPS-43 SMGs during the German retreat from Stalingrad. Soviets made good use of their SMGs as a result of their stunning losses in the 1940 Russo-Finnish War.

The invention of the Sturmgewehr, or assault rifle, caused arms experts to predict the demise of the SMG for the past four decades. So far, however, it is still popular.

The third generation SMG demonstrates the desire to reduce the overall size of the weapon by using a telescoping bolt. This allows the barrel to be mounted well inside the receiver instead of at its end. Shown is the Czech Vz-25.

began at a fevered pitch—everywhere except in the United States, which had already reverted to pre-war attitudes about small arms development.

When communist troops rolled across the 38th parallel in 1950, another, perhaps more significant kind of war was introduced. It typified the "brush-fire" concept of irregular, dirty warfare that was to become common in years following. Close range, massed attacks were the enemy's rule and the SMG's firepower was sorely missed by U.S. and Allied troops. The communists, however, were liberally armed with PPSh-41 and PPS-43 SMGs, as well as large quantities of Thompsons, and enjoyed a significant advantage when these arms were used in conjunction with large-scale stealth tactics prior to the actual assault. That the Thompson was considered a prime war trophy by U.S. troops and often used in preference to their own issued arms speaks for itself.

The 1950s and 60s saw continued prediction of the SMG's replacement by the assault rifle, but at the same time also saw the SMG's further growth and proliferation. Among these advances was the "third generation" SMG, typified by the Czech Vz-25 and its more famous counterpart, the Israeli UZI. Few arms have achieved the notoriety of the UZI. The last time I checked the sales figures, the world was gobbling them up from its producers at Fabrique National and Israeli Military Industries as fast as they could be manufactured.

Gordon Ingram, long prominent in the field of SMG design, produced in the late 1960s his M-10 and M-11 series of third generation SMG. Both have also gained worldwide popularity and utilization, particularly as special-purpose weapons used with a sound suppressor.

The 1970s saw a total shift of the communist bloc to the AK assault rifle for logistical reasons, but the SMG continues to live on in the heat of a thousand battles, large and small. The UZI fought the 1973 Yom Kippur War, as did the Egyptian version of the Gustav M-45, the "Port Said." The Thompson, MP-40, STEN, Sterling, Madsen, and others continued to be heard in Asia, Africa and Latin America, used against Soviet PPSh-41 and PPS-43s in the hands of the opposition.

Another trend of the 1970s was the small arms "package deal" in which several weapons designs were offered for sale in quantity by a single manufacturer. Pioneered by Heckler & Koch, considerable commercial advantage is gained by such a method and, although the weapons involved in any such "system" may or may not be the best available, the inherent convenience of such an offering to the potential buyer cannot be denied. H&K's MP-5 SMG typifies this concept in that it is part of such a system and reflects all of the operational characteristics of the other arms within it. Technically speaking, the MP-5 is overly complex and expensive for SMG applications due to the fact that it fires from a locked breech. This, in turn, allows heat to build up quite quickly and cause considerable discomfort to the firer as well as reduced service life of the weapon. Regardless, the MP-5, a true "fourth generation" SMG, has found favor with many elite counterterrorist units around the world because its closed bolt mode of fire allows precise head shots against terrorists holding hostages as shields. The fact that the gun tends to overheat is negated by the fact that it is rarely used in a sustained-action role during such operations.

The technological advance of the SMG stopped with the third generation, which utilized a "telescoping bolt" to allow drastic reductions in the weapon's overall size and length, and it is at this point that it remains today. As this is written, the experts are still expounding on the imminent death of the SMG and I think that, sooner or later, their prediction must come true. On the other hand, I am also a realist, and I know the SMG has remained healthy in spite of everything else. The reason is simple: it fulfills an important practical mission which has been left vacant by the assault rifle. It is easy to use effectively and economical

The Egyptian "Port Said" SMG squared off against the Israeli UZI during both the 1967 Six-Day War and the 1973 Yom Kippur War.

The author is shown here with his UZI with suppressor.

Still alive and well after more than sixty years of combat, the SMG continues to serve. Shown here are two British Royal Marines with their Sterling L-2A3s during the recent Falkland Islands campaign.

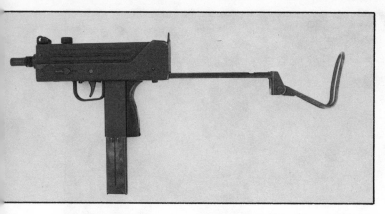

The Ingram M-10, an example of a third generation SMG.

The Heckler & Koch MP-5A3 SD, most popular example of a fourth generation SMG.

to produce. It is small and light and is completely satisfactory for so-called modern combat. Do you doubt this? If so, I ask that you remember that the typical engagement range worldwide is less than twenty-five meters and while this sometimes shocks the academician into stunned disbelief, it is common knowledge among those "who have been there." The late General S.L.A. Marshall, that supreme military historian, repeatedly attempted with little success to make this point understood. He even went so far as to describe the phenomenon as it occurred in even the open spaces of World War II and Korea and emphasized that point-black ranges were by no means unique to Vietnam. In his excellent books, *Night Drop, An Evaluation of U.S. Infantry Weapons and Tactics of the Korean War* and *Ambush and Bird,* Marshall outlined with terrifying detail the scenario in question, even though he is talking about action in three different wars!

This, then, is why the SMG doesn't die, because this kind of fast, close-in action is precisely the mission for which its ease of handling and portability are suited. If it is indeed a "slob's weapon" as espoused by one firearms writer, then all who participate in such combat are "slobs." This is clearly not the case.

The modern "combat" shotgun has been adorned with phenolic pistol-grip buttstocks, muzzle brakes and folding stocks. With this I have no quarrel since phenolics are considerably more robust than wood. It is my personal feeling that muzzle brakes are a waste of time on a weapon of this type because Newton's Third Law of Motion governs recoil and cannot be significantly altered by a direction of low pressure gases. The folding stock has merit if storage space is a problem. But only then, because one sacrifices critical handling

characteristics in return for it. To be a worthwhile tradeoff, an actual need for it must be apparent.

In conclusion, the selection of either the SGN or SMG must be based upon an evaluation of need. One cannot efficiently use a screwdriver to drive nails, although he might make do if required. In much the same way, a shotgun is not the tool for action beyond fifteen meters or in heavily wooded environs. Nor is the SMG the answer to a home defense problem, especially if there are innocent people present in the home who might be endangered by over-penetration. The shotgun is socially and politically acceptable for general law enforcement work whereas the SMG is not because it presents a heavy-handed image that is intolerable to the populace. Like it or not, stereotypes, even false ones, govern our lives far more than we realize.

Yet for SWAT team operations, the same news media who chastise the SMG for general police use expect SWAT personnel to carry exotic weapons as a matter of policy. In this arena, the SMG is experiencing great proliferation and popularity, especially among entry teams. SWAT operations are typified by fast urban indoor action—the perfect mission for a properly employed SMG and one for which the shotgun is at best marginal due to its bulk, excessive recoil and tactical inflexibility.

Choose your weapon carefully and learn to use it well. Be capable of working with it efficiently—rain or shine, night or day. Be aware of its strengths and limitations and use it accordingly. Don't let your ego and imagination run away with you because such things have no place in the life and death world of which we speak.

If you follow these directions, you can't lose, regardless of which weapon you select—because you will have chosen the "right tool for the job."

2. A PROFESSIONAL'S LOOK AT THE SHOTGUN AND SUBMACHINE GUN

I have for quite a few years worked in situations involving the SGN and SMG in both training and actual operations. In this vocation, I have travelled worldwide and seen on three continents the way combat action begins, develops, and concludes.

Naturally, anyone who has experienced these things cannot help but form some definite impressions, based upon firsthand observation and field participation, of how the inherent characteristics of both weapons stack up, not against each other, but on an individual basis.

In the section that follows, I have placed in photographic essay form with my impressions and observations, these thoughts for your perusal and consideration. If you disagree with my opinions, so be it. I ask only that you consider them and why they were stated in the first place, remembering at all times that this book and my profession deals strictly and specifically with the efficient use of these weapons in real situations.

I am certain that for hybrid utilizations such as IPSC competition some of the negative impressions I have about certain weapon characteristics might not apply. This, however, is outside the realm of my vocation and I make no further comment upon it.

Take a long look at what I have set forth here before you form any opinions of your own about what constitutes a "good" fighting shotgun. I'll tell you right now—up front—that I am a staunch believer in the KISS (Keep It Simple Stupid) principle and use it along with Murphy's Law in the administration of ALL of my training and operational activities. Inasmuch as I have survived (and so' have my students and associates), I feel safe in claiming their validity. It is my belief that if you consider the pros and cons of reality versus commercialism, you will also.

I have divided this section into two parts; one dealing with the combat shotgun and the other with the submachine gun. You will find no comparisons of characteristics unique to one weapon placed against the other, for, as I hope you are by now well aware, I am convinced that the two arms share no real common bond of application or concept other than the fact that they are both firearms. Instead, I have compared the most common features found on various makes of shotgun and SMG. Placed in such a format the information is both more relevant and easy to understand.

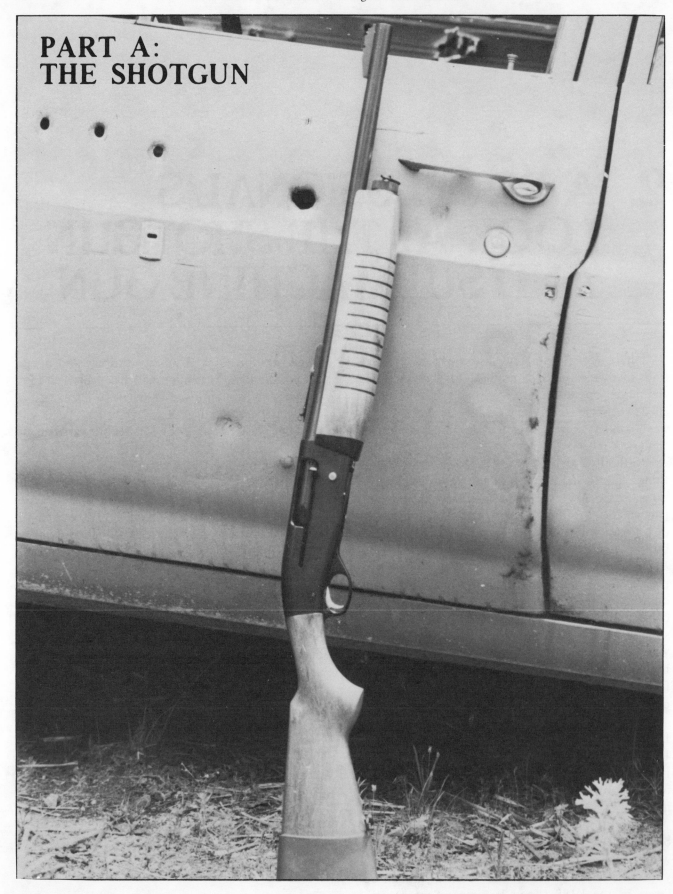

PART A: THE SHOTGUN

The SKB Riot Gun.

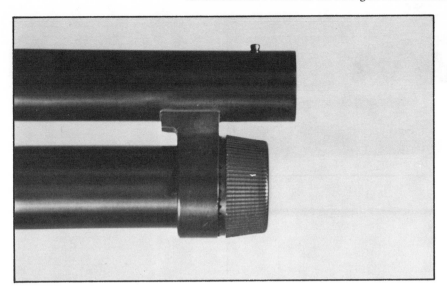

There is considerable difference in opinion as to the utility of the various front sights found on shotguns. Since buckshot is the ammunition primarily used in the weapon, the standard brass bead is fine as long as it is large enough to be seen quickly under stress.

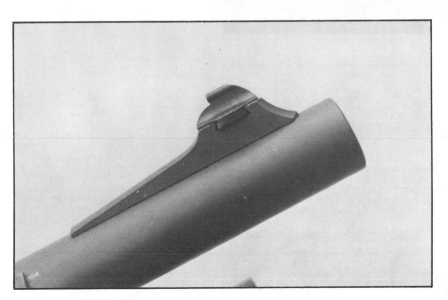

Rifle sights on the SGN (shown in the center photo, left, and on the following page) can be as much of a weakness as they are an asset. Since slugs are used to hunt deer in highly populated areas, many shotguns are provided with sights applicable to such employment. This does not mean that your fighting SGN must have them. They increase the height of the gun and are fragile and thus easily damaged.

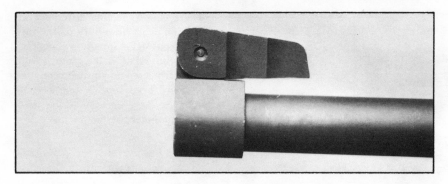

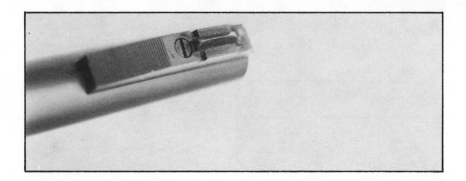

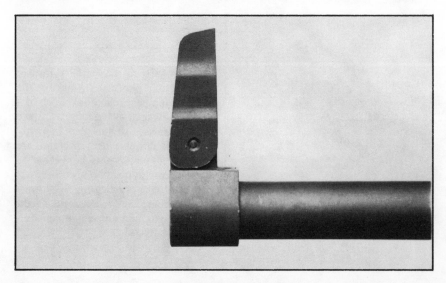

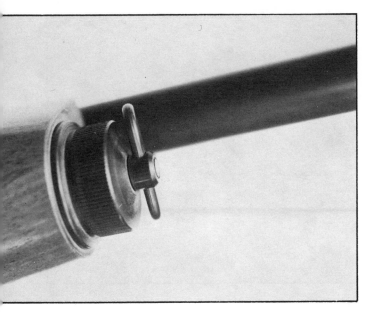

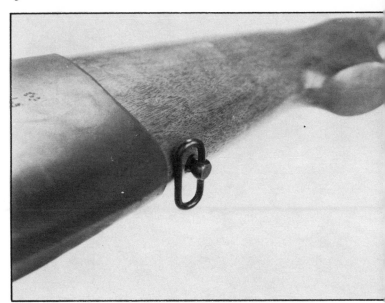

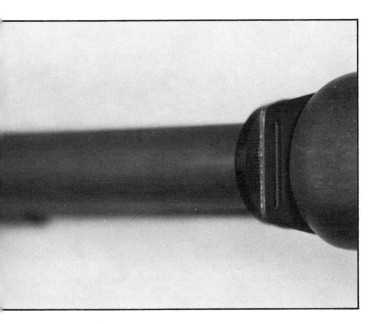

The way in which sling swivels are mounted on the SGN can make considerable differences in utility. Since shotguns are rarely used for long-term activities in which a sling is necessary, the author feels that the swivels can be just as easily left off the weapon entirely. If, however, you wish to use them, be certain that they are solidly mounted and placed where they will not interfere with efficient use of the gun.

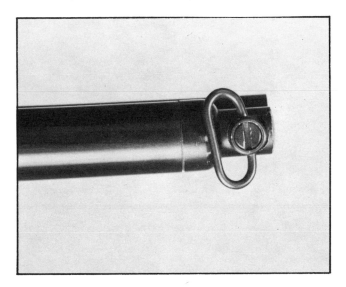

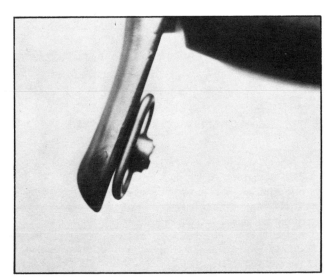

Magazine tubes (top and right) often loosen during carry and firing if they are not properly secured. Most attachment devices tend to loosen over a period of time and require constant inspection to prevent loss and functioning problems with the gun. Sometimes a dash of Loctite is in order.

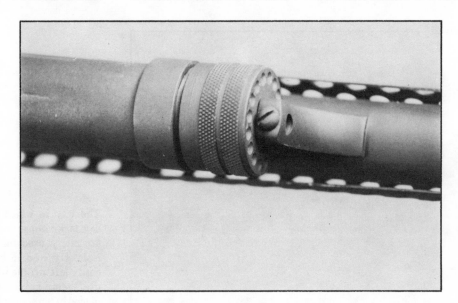

The actuators (right and top of opposite page) of most self-loading SGNs are somewhat awkward. Some are sharp, causing abrasion to both skin and clothing. They should be inspected for burrs and "dehorned" as required.

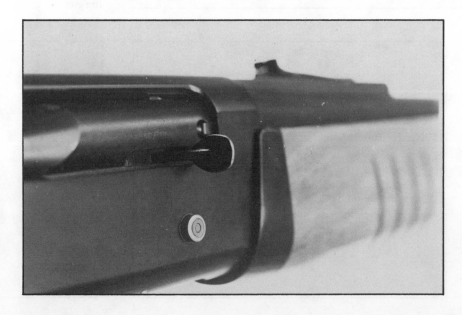

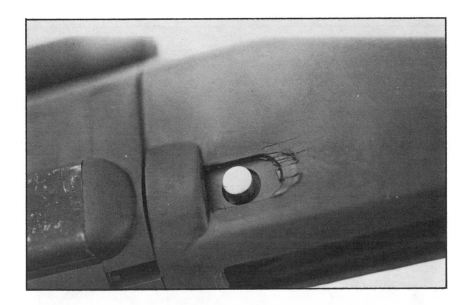

Semiauto shotguns require manipulation of an action-release button (center and left) before they can be loaded. This device should be deactivated as quickly as possible to allow quick, efficient loading under stress.

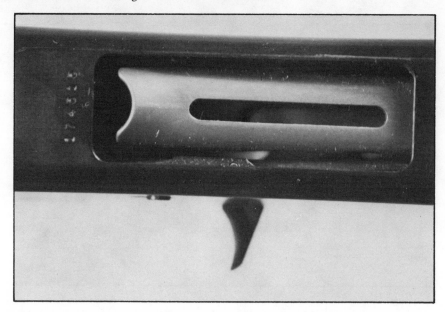

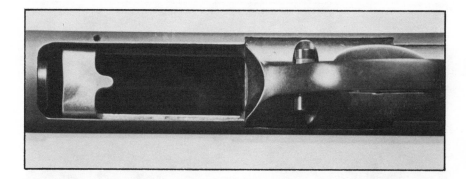

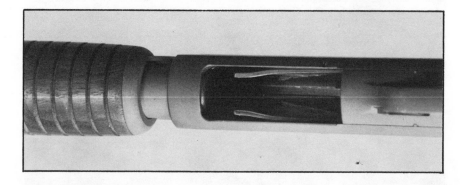

One of the inherent problems with SGNs utilizing tubular magazines is that occasionally a shell will jump the shellstop and lodge beneath the lifter and bolt, making cycling of the action and ejection of fired shells difficult and sometimes impossible. Some guns come from the factory with "skeletonized" lifters to allow access to the shell during such situations (top three photos). If yours is not skeletonized (photo right), then modify it yourself.

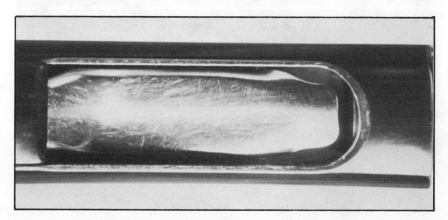

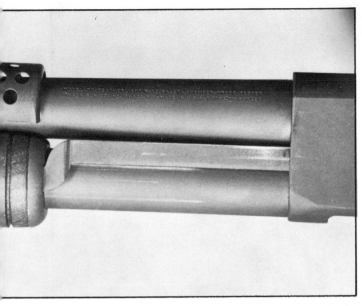

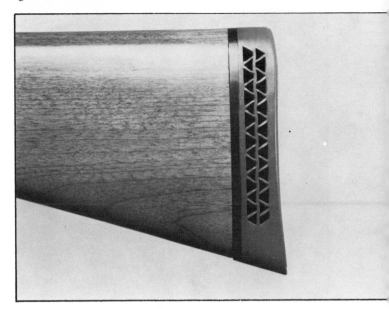

Some feel that the slide-action is stronger and more reliable with two action bars instead of one. However, inasmuch as the most famous shotguns have gotten along without them with no difficulty for more than seventy years, a single will accomplish its function quite satisfactorily.

The recoil of the shotgun bothers a great many who would otherwise opt for its use. The addition of a recoil pad is a good idea and in no way reduces the effectiveness of the gun.

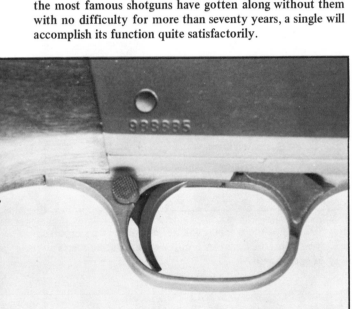

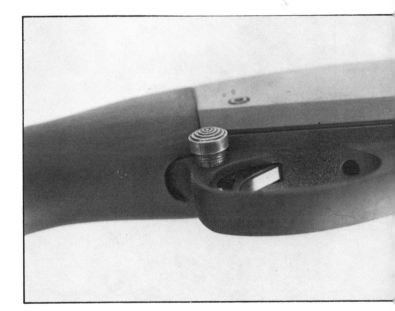

Other than its location, which is often too far forward for efficient manipulation by the trigger finger, the shotgun safety button (above) is too small and should be modified or replaced with a larger one (right). The action release is also often poorly placed and too small.

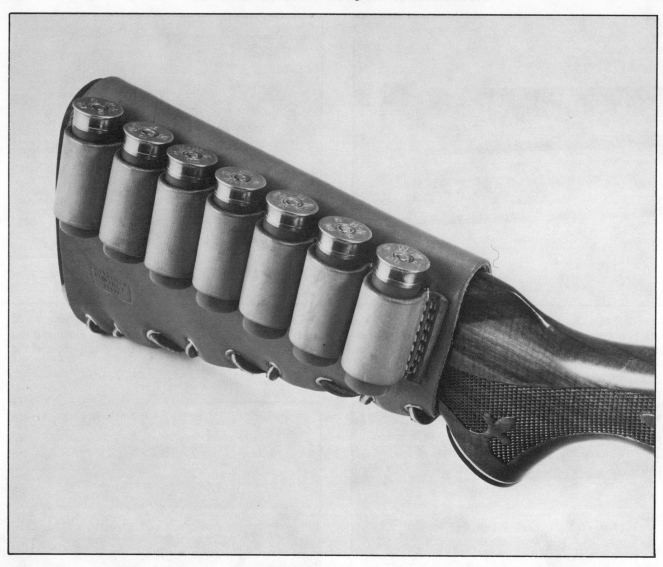

Buttstock-mounted ammunition carriers prevent efficient underarm utilization of the weapon and detract from the weapon's balance and handling abilities. Moreover, a carrier is largely unnecessary as shotgun scenarios requiring spare ammo are typically too abusive on the weapon and operator to permit the open carry of spare shells.

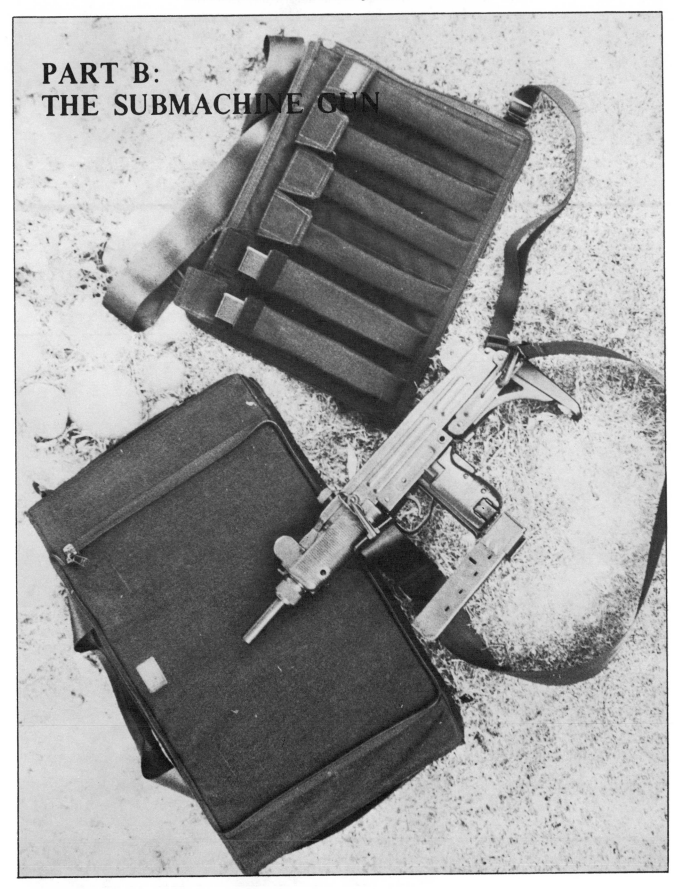

PART B:
THE SUBMACHINE GUN

The 9 mm UZI.

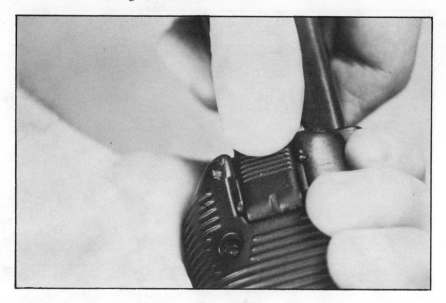

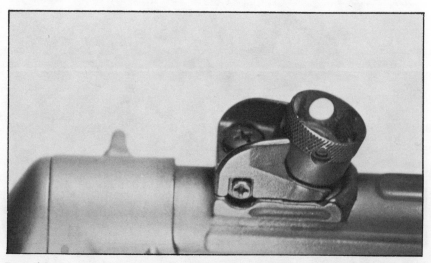

Rear (above) and front (opposite page) sights on SMGs are as varied and often as confusing as those on shotguns. However, they are generally more robust and are well designed for sustained abuse.

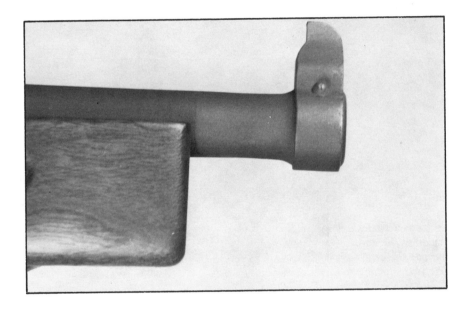

The Heckler & Koch MP-5 protective ring (top) actually interferes substantially with quick sight acquisition and requires modification as shown (right) to alleviate the problem. This should be rectified by the manufacturer, not the purchaser.

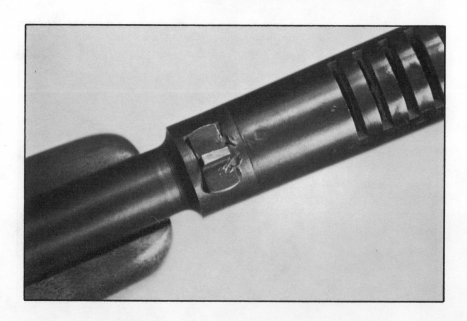

The method of attachment (peening) on this Reising M-50 front sight is totally unsatisfactory and conducive to loss of zero.

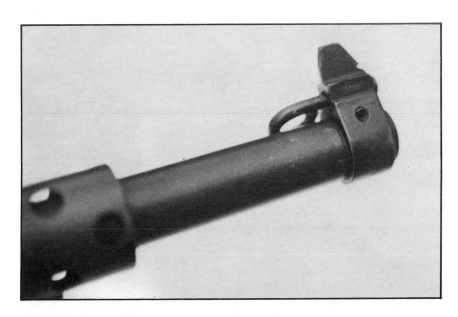

Sling swivels (all photos on this page) are perhaps of more use on an SMG than on a shotgun but watch out for the rattle most of them produce. It can get you killed. A bit of tape can cure the problem quickly.

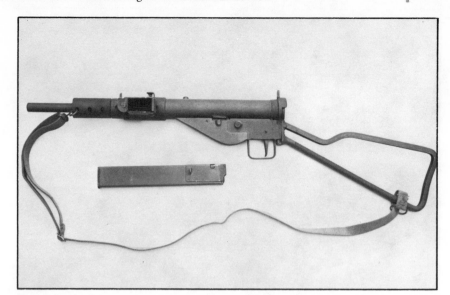

Sling-equipped Mark II STEN. If properly mounted, a sling can increase stability when correct technique is employed.

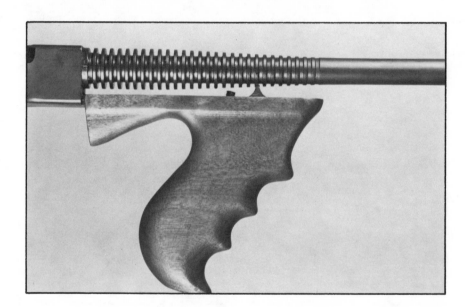

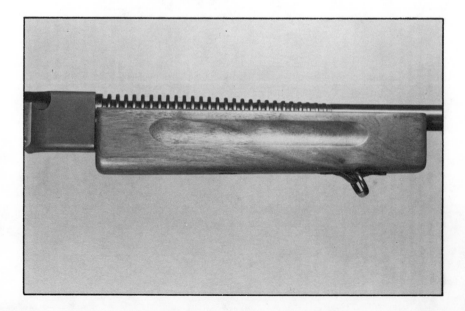

Fore ends of typical SMGs are of either horizontal (right) or vertical (center) configuration. Either may be constructed of wood or of metal. Of the two, the horizontal is by far the most efficient.

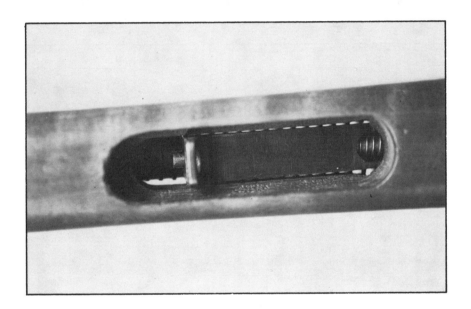

Actuators or cocking knobs (all photos on this page) can also be found in many different shapes and locations. All work well as long as sharp edges are removed from them. The author prefers those which have been mounted for easy manipulation in combat.

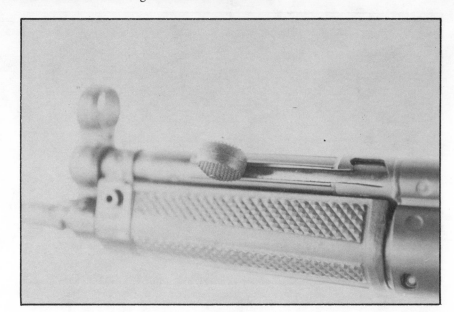

Care must be taken not to damage knobs that are made of plastic, such as on this H&K MP-5. If the knob shatters, and it often does in sustained field use, it becomes both difficult to operate and sharp.

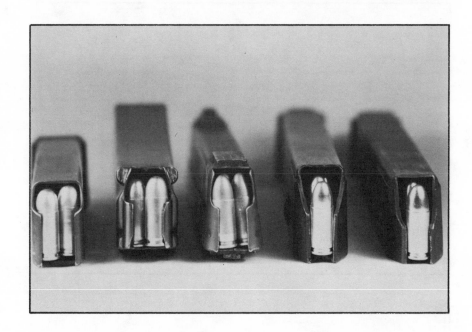

Both double and single column magazines are used in SMGs. The single-column types normally move the ammunition from a double column into a single just before it reaches the feed lips. The author prefers the double column for ease of loading and reliability. However, we don't often have a choice!

As with any other arm that utilizes a detachable magazine, inspection on a regular basis is required to ascertain field serviceability and reduce the likelihood of stoppages in combat. During inspection, the feed lips of a detachable magazine, as well as the magazine body itself, should be scrutinized closely for damage.

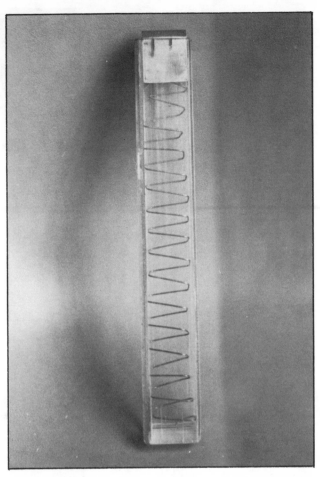

Modern technology has yet to perfect a plastic magazine that can withstand field treatment and still remain functional, although theoretically such is possible.

Drum magazines (shown in right of photo) are relics from the first-generation SMGs. More streamlined box magazines have been preferred since about 1940.

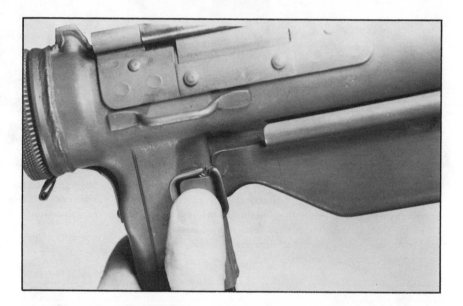

Magazine release devices are normally well placed on most SMGs (all photos on this page), allowing fast magazine changes with little effort.

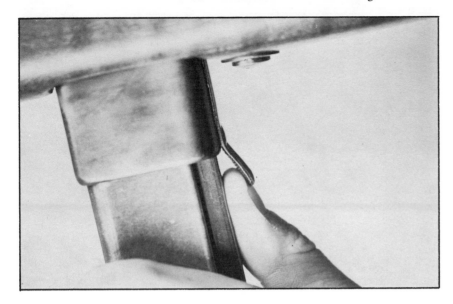

The Reising M-50/55 requires a strong rearward pull of the release to allow the magazine to come free of its housing. This is poor human engineering.

The H&K MP-5 features a dual release with both a pushbutton and tab. While perhaps unnecessary, there is no harm done and it certainly is convenient!

On second generation SMGs, a bolt lock is a good idea to prevent accidental discharge if the weapon is dropped directly on its butt. Shown are the PPD-34 (top), STEN (center), and MP-40 (right).

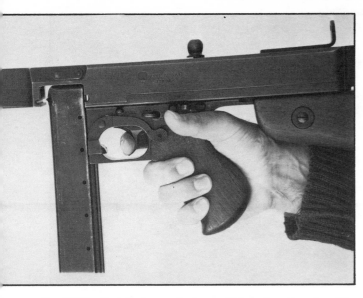

On SMGs that feature selector switches, location is critically important. Most first generation guns have poorly located selectors, such as this M1928 Thompson.

An example of selector switch placement that is an afterthought.

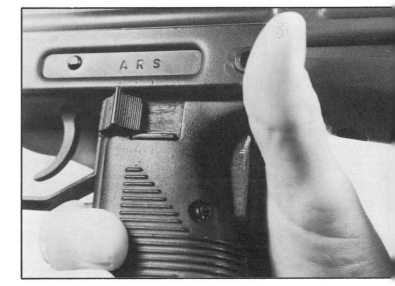

Although often too short for most operators, a combination safety/selector switch is a good idea (above and right).

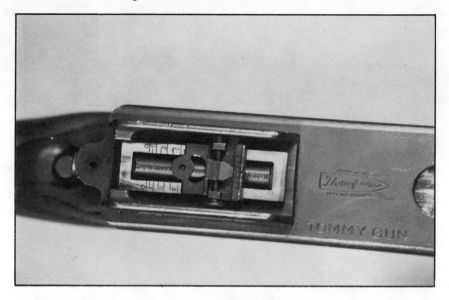

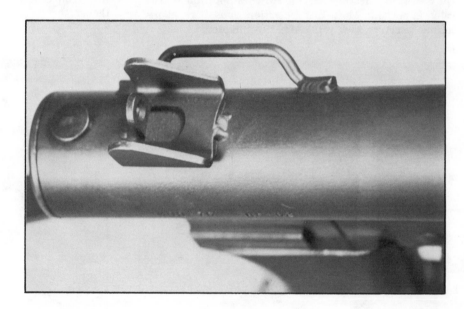

Rear sights can also have an adjustable or fixed U-notch.

Rear sights for the SMG typically take the form of an adjustable or fixed aperture (photos above and on preceding page).

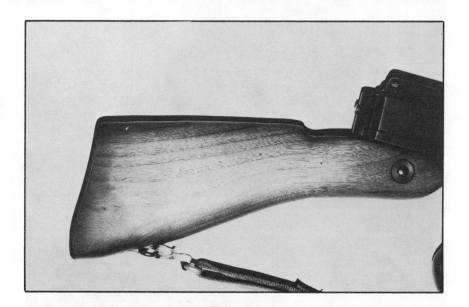

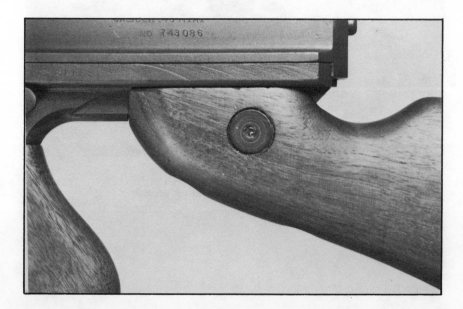

Buttstock designs on an SMG are as varied as every other feature (photos on this and the following page). They can be made in virtually any shape or design, fixed or removable. Most second and third generation guns feature a folding or telescoping stock, as do all fourth generation models.

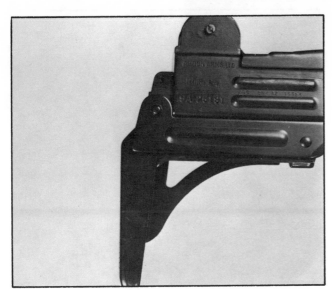

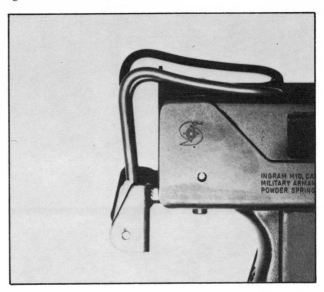

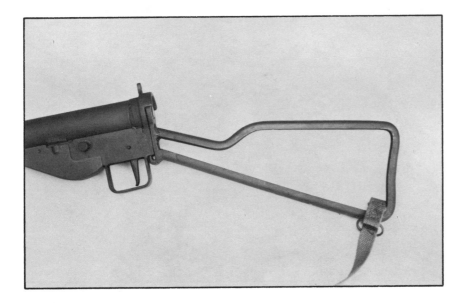

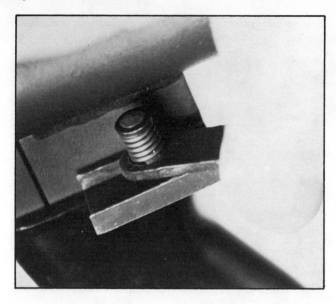

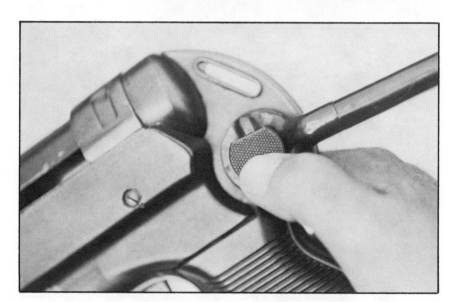

One valid complaint about most folding SMG stocks (right and above) is that they quickly become loose, allowing excessive wobble of the gun when shouldered.

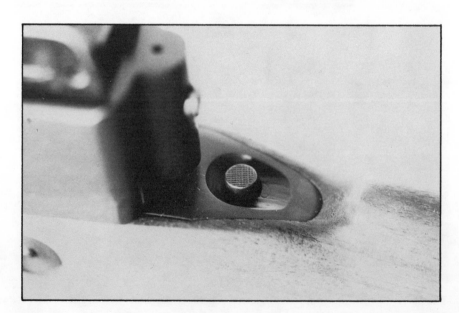

Shown is the Thompson SMG with removable wooden buttstock. It is solid and effective, if a bit expensive.

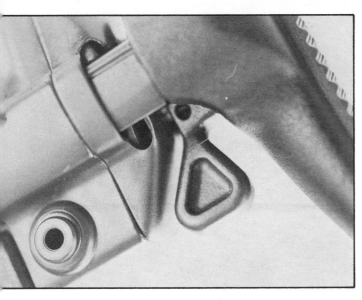

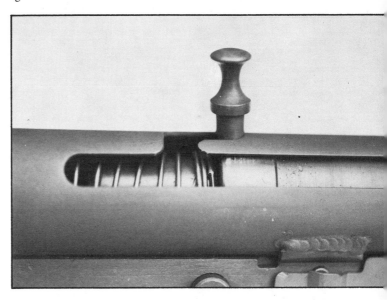

Heckler & Koch retractable stocks lock up solidly without wobble. Unfortunately, they are often too short for most firers, a problem shared by many other designs.

Several second generation SMGs feature a simple notch as a safety. Although rudimentary, these function surprisingly well.

The MAT-49 of Indochina fame featured a folding magazine housing to further reduce the weapon's bulk while being stored or carried. This is an excellent idea.

3. WEAPON AND AMMUNITION PERFORMANCE

As eloquently stated in Eric Strahl's forward, there are no other two small arms which suffer from more misconceptions, legend, and just plain poppycock when it comes to performance. I have been able to isolate several causes for this, including simple ignorance, cinematography, ambition, and just plain dishonesty.

Earlier in this work, the point was made that the shotgun's early popularity was in large part assisted by its graphically simple concept; in other words, what it does and how are relatively easy for the uninformed mind to grasp. While this remains true today, the passage of time takes its toll on everything—including the facts—twisting and reshaping them into a form which often bears little resemblance to the truth.

Such has been the case with both the SGN and, to a slightly lesser degree, the SMG. In the instance of the shotgun, I suspect that the fact it has been around in one form or another for so long has had as much influence as anything else. For decades, myths based upon little more than war stories told over a beer in the local tavern have been perpetuated, magnifying each weapon's effectiveness while downplaying its limitations and true performance. If this goes on long enough—and it has in the case of both arms—legend takes over; the attitude that transcends all fact, logic and reality. This is especially dangerous because those barroom tales are based as much on egotism as they are on fact. When the element of telling a good story is added into the mixture, it is easy to see how things get out of hand.

I doubt that any reader of this material is surprised by this. However, what *is* surprising is that these same myths are restated continually by people who claim to be experts with these guns as proof of either weapon's superiority. I find this trend disappointing because, as a result, precious little, has been done to determine what really happens when the SGN and SMG are utilized in the real world.

This situation is dangerous because of its potential effect on anyone who uses either firearm for serious business. I see it all the time, even in law enforcement training programs, where the myths about the shotgun's effectiveness and versatility run so rampant that they are actually frightening to anyone who knows better. On the opposite side of the coin we have similar problems with the SMG, but normally these distortions are negative; that is, how *in*effective it is, especially as compared to the SGN. Again, the culprits are egotism, simple ignorance, ambition, or actual dishonesty. In every case where I have witnessed such goings-on, the above were quite obvious. Moreover, the "expert" delivering the sermon was usually at least adequately competent with some other form of firearm, thus preserving the facade of his expertise.

I, too, watch movies and television. Indeed, I have participated in script consultation, technical advising, and the actual training of cinema personalities, but I caution the reader about taking what he sees on the silver screen too seriously. Successful movies and TV shows involving guns all exhibit one common element—action. Action is emphasized by sensationalism; spectacular displays that dominate the senses. In order to create these displays, myths and romantic notions are exploited by "creative" camera work and elaborate stunts. We've all seen them; the scenes where the marshall blows a bad guy literally through the wall of the saloon with his shotgun, or where a war hero guns down dozens of enemy soldiers from the hip with his SMG at ranges that stupefy us—and never once lets up on the trigger, the gun streaming an endless supply of empty cases from its ejection port.

Do you wonder where the misconceptions come from? We have all experienced some degree of subliminal indoctrination with these erroneous concepts because we see them so often that our subconscious minds begin to accept them even if our intellects do not—and therein the danger lies. This process is further accomplished by incorrect statements made by people who *think* they know what they are talking about. Guess who else has been to the movies!

I have seen this phenomenon on several occasions, especially with the pistolero or trap and skeet shooter who suddenly becomes an expert with the "combat" shotgun. The two categories have nothing whatever in common—the techniques, ammunition and guns are as different as night and day. Inasmuch as this scenario occurs surprisingly often, I again caution the reader to be careful about what he accepts from anyone. Ask for supporting reasons and don't be misled by off-the-cuff answers. An expert will have the answers and can give them to you without becoming abusive or evasive. Any *real* professional can listen patiently to students' questions and then give them intelligent answers, with supporting facts. When you witness a "put down" or tantrum, watch out!

Through my association with SWAT magazine, I am fortunate to have been provided with the resources to actually determine what the SGN and SMG, among other arms, can and cannot accomplish, through actual field testing under both controlled and uncontrolled conditions. Some of the results of these experiments have been surprising, others not at all. The tests were realistic and the conditions present during the tests were representative. Remember that what you will see on the following pages illustrates a norm, a yardstick for what you can expect, *not* an absolute, for no such thing exists.

My own specific impressions of these tests deal with the misconception that the shotgun is a versatile weapon, which in combat it is not. Being able to utilize many different shot types as well as slugs makes it versatile as a gun used to hunt everything from upland game to migratory fowl to deer in highly populated areas where a rifle is impractical or illegal. On the other hand, this in no way makes it versatile in combat, for there are only a few choices of shot, specifically No. 000, 00, 0, 1 and 4 buckshot. This already limited selection is made smaller when one discovers that only rarely does an individual shotgun pattern adequately with more than one of these, if at all.

Too, the idea that one can change rapidly from buckshot to slugs to engage targets at moderate ranges comes from ignorance and overattention to theoretical possibility. In fact, especially under stress, the firer cannot usually accomplish this feat, the typical result being that he cannot even remember where he keeps those precious slugs at the moment of action. This is not an unsupported opinion; it is based upon actual testing of personnel whose skill levels ranged from novice to expert.

The subject of slug accuracy is also a sore point. In my tests, I found that, other than being highly visible, slugs are virtually worthless beyond about fifty meters because their stability deteriorates rapidly. An "accurate" slug shotgun will produce three-shot groups at fifty meters that range from three to five inches. But at seventy-five meters that same gun often misses the target entirely! Slugs don't penetrate as well as we believe, either.

Turning back to the matter of buckshot, the *SWAT* magazine tests turned up some interesting facts. First, buckshot is a marginal penetrator. Second, it is worthless against automobiles. Third, it is not nearly as effective in light vegetation cover as previously thought, especially when the target is some distance *beyond* the intervening cover. Fourth, a review of case studies involving shotguns with buckshot indicates that shotguns tend to be used most often in handgun-type situations, perhaps the result of their widespread law enforcement use. Fifth, buckshot dispersion occurs so quickly and erratically that the maximum effective range of the SGN, as a general rule, rarely reaches twenty-five meters. Sixth, from a review of number 5 above, shotgun effectiveness against unarmored targets in the open revolves around how long one can keep the shot pattern tightly configured, *not how quickly it spreads out.* This totally destroys the myth of the "alley sweeper." Shotgun pellets are not the equivalent of a pistol or SMG bullet and their soft composition, poor ballistic shape and general lack of consistency do much to limit their efficiency.

With respect to the performance of the SMG, I note that:

1. SMGs penetrate windshield glass and body-work quite satisfactorily as long as the angle of bullet entry is not too shallow.

2. SMGs perform quite well in light cover regardless of whether the target is located directly behind the cover or some distance beyond it.

3. SMG ammunition, even though it is of pistol caliber, performs noticeably better from an SMG than it does from a handgun.

4. Effective SMG ranges can, with few exceptions, exceed 100 meters when proper technique is employed.

5. There is no evidence to indicate that a multiple-shot burst from an SMG is any less effective than a shotgun blast or hits from other types of small arms.

6. The SMG is quite easily used well by people of small physical stature due to the weapon's low recoil.

7. There are only a few instances when automatic fire should be employed from the shoulder.

8. Underarm assault fire, the natural home of full-auto, should not be employed beyond a range of ten meters from the target. In such situations, it is actually faster to shoulder the weapon and use the sights to place single shots on the target.

9. SMG versatility far exceeds that of the SGN. That the SMG has long been regarded otherwise stems from the widespread myth that it is supposed to be fired "from the hip" in long, indiscriminate bursts.

10. The common notion that IPSC-type targets, typically utilizing a 10-inch/25cm "dinner plate" type X-ring are an accurate representation of a vital area hit, is completely erroneous. Such targets are invalid as a measurement of either weapon performance or operator skill. This particular observation applies to the SGN as well.

11. The SMG is most effectively employed in the full-auto mode using two-shot bursts if point targets are engaged.

12. Bursts of fire exceeding two shots are most effective against small area targets or tightly grouped multiple targets at close range.

13. The SMG is highly effective in SWAT and counterterrorist missions. It is precise enough to handle delicate situations involving hostages at close range, but can also successfully engage opponents under a wide range of tactical conditions.

I urge the reader to examine this section carefully several times to gain an understanding of the points outlined. If this is accomplished, a good perspective on this critically important aspect of the two weapons' capabilities will be achieved.

PART A: THE SHOTGUN

The myth of SGN versatility is shattered when one realizes that for combat applications there are only a few shot loads available and that any given shotgun will pattern adequately with only one or two.

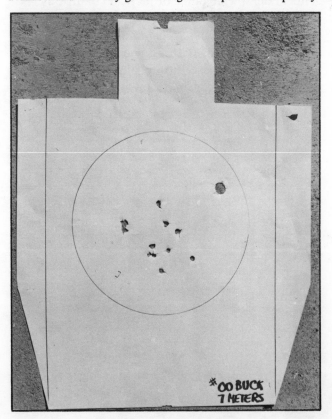

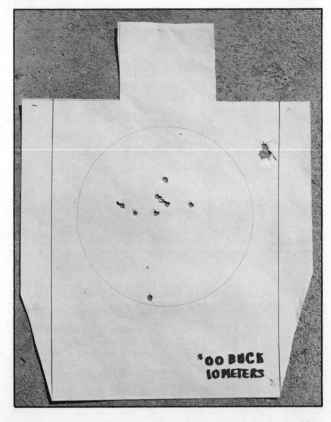

This is the best SGN patterning I have ever seen. Note that while pattern density against a human adversary is quite satisfactory even to 25 meters (a rarity among shotguns), its effectiveness completely disintegrates beyond that point. This problem

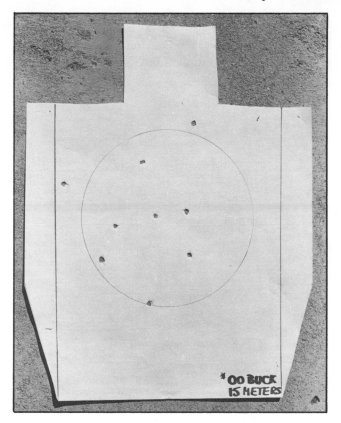

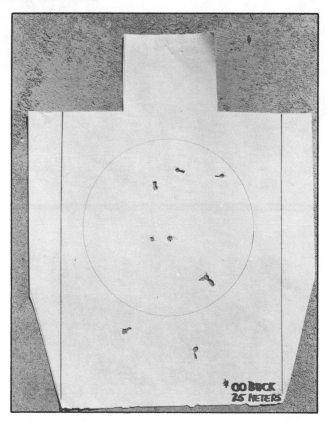

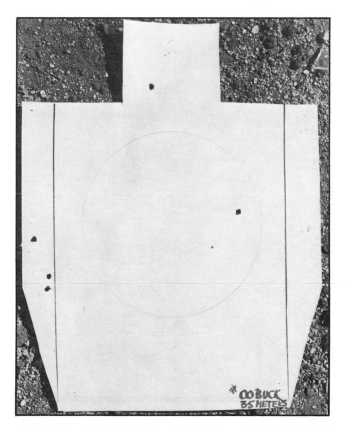

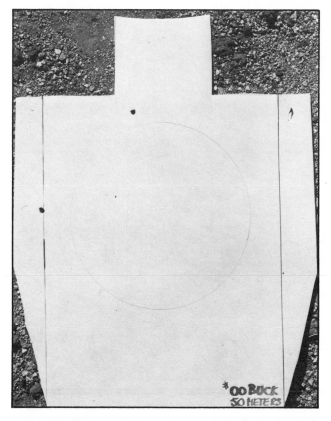

causes many novice shotgunners to select the smaller sizes of buckshot because the larger quantity of pellets give better patterns. The problem here is that smaller shot sizes mean lower penetration, already a weak point of the SGN.

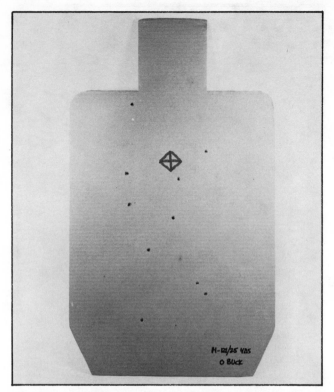 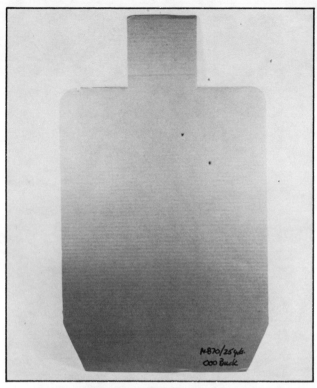

Shotguns are highly variable in their patterning performance, even at identical ranges. Shown here are two targets, the one on the left shot with 0 buckshot from a Benelli M121 at 25 yards, the one on the right with a Remington M-870 with 000 buck. These represent two opposite ends of the performance spectrum and illustrate how erratic SGNs can be. Careful testing for patterns vs. penetration capability can do much to reduce this problem.

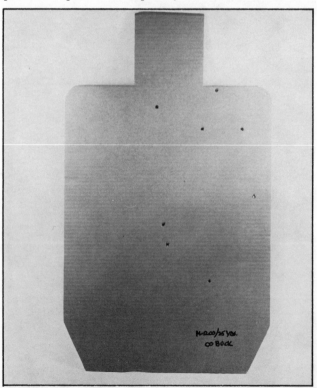 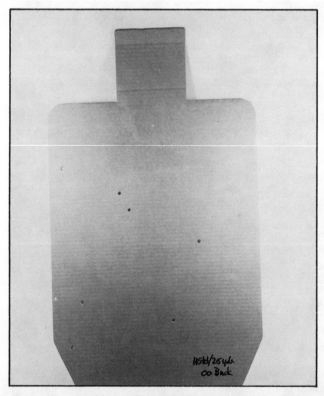

Beginning at about 15 meters, it becomes evident that many SGNs do not shoot to point of aim with buckshot. At left is a pattern from a Winchester M-1200 with 00 buckshot at 25 yards that went to the upper right and would require a low/left of center hold. At right is a target shot with High Standard 8110 with same load showing low/left impact requiring opposite hold. Only patterning done during training before actual use of the weapon can disclose these important tendencies.

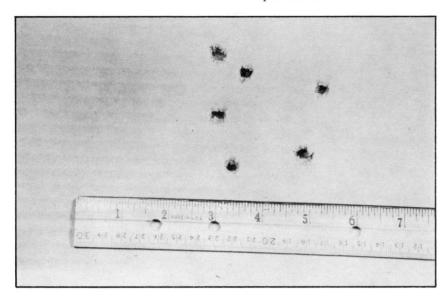

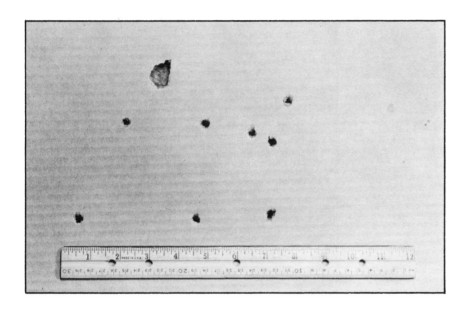

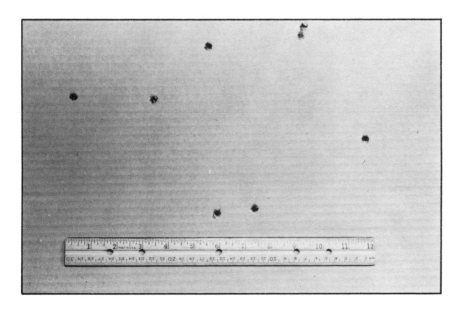

In direct contradiction of the popular myth that maximum shot spread is desirable, reality dictates that the ideal is to hold the pattern together as long as possible. SGN pattern deterioration occurs rapidly at close range as it is. Here are shot patterns with 00 buckshot fired at 5 meters (top), 7 meters (center), and 10 meters (left). Note the increase in pattern size even at these ranges.

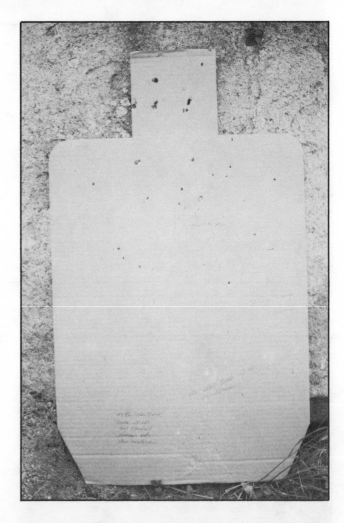

Buckshot fired from 18- and 20-inch barrelled SGNs does not perform well against automobiles. A target (bottom) hit through a windshield (top) with 00 buck from a Remington M-870 riot gun from 10 meters showed no penetration and little damage except by small fragments from the inside layer of safety glass.

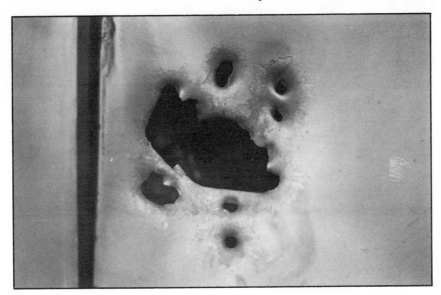

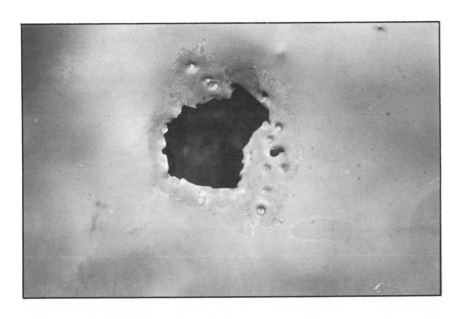

Typical automobile bodywork offers excellent protection against all sizes of buckshot. Shown here are door panels shot from 3 meters using an Ithaca M-37 with a 20-inch barrel with #1 buck (top), 00 buck (center), and #4 buck (left). None penetrated the interior of the vehicle.

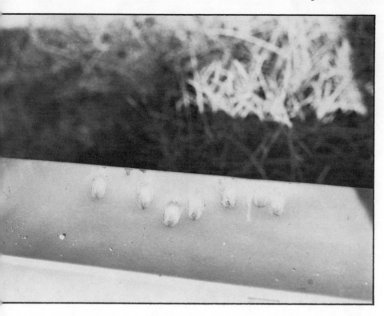

Buckshot also deflects easily from sheet metal when fired from shallow angles. If this characteristic is carefully used, it can be turned into an effective technique against careless opponents.

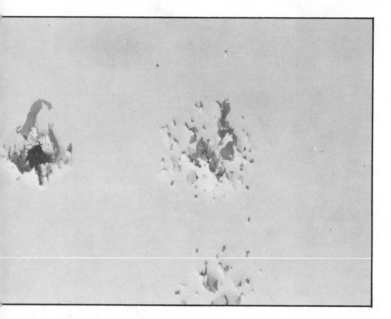

Typical sheetrock wall panel offers little resistance to buck-shot, requiring the use of light, small shot loads for urban home defense where innocent persons are present. Photo shows #8 trap loads against sheetrock as fired from a Winchester M-12 with an 18-inch barrel. Lower velocities of shot from a shorter barrel limit penetration to acceptable levels.

Shotgun slugs are largely overrated in terms of both accuracy and effectiveness. Typical slugs used worldwide are (from left to right) Brenneke, sabot, and the standard U.S. type.

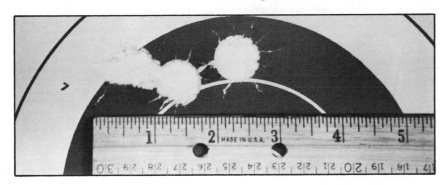

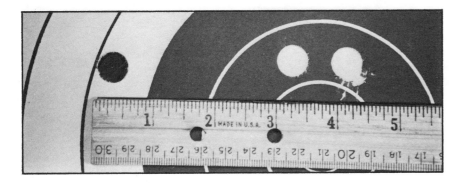

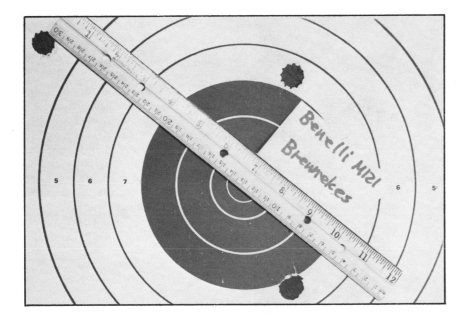

Slug accuracy varies drastically even with the same gun. As with buckshot, careful testing must be done in advance to determine what load your particular SGN likes best—or if it will perform adequately at all. Top photo shows a target hit with a Benelli M-121 with standard U.S. slugs. Note that one shot, although still in a group, is already beginning to tumble. Center photo shows sabot slugs. Here, too, one is beginning to lose stability. Fifty meters, the range at which these groups were shot, represents the maximum at which we can expect reliable performance from slugs. Although Brenneke slugs are the best penetrating (left) of the slug types, it is consistently the least accurate.

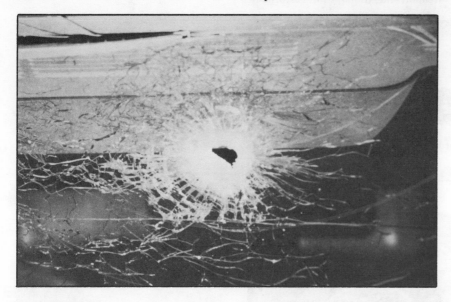

Slug performance against automobiles was shockingly poor, considering the claims made for them by many law enforcement instructors during training lectures. Typically, slugs tend to fragment on windshield glass (top) and at best pepper the target with fragments (left). Whether or not the wounds inflicted would be incapacitating is problematic and subject to doubt.

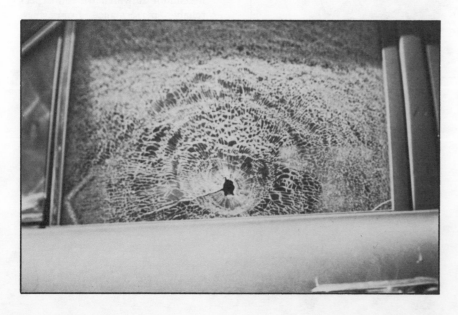

The only place where slugs performed well was through side glass. No problem here!

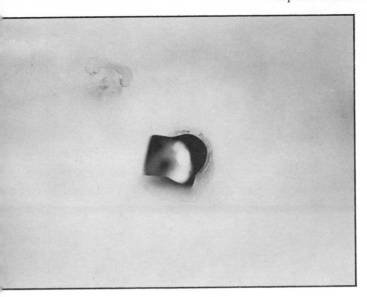

A U.S.-type slug fired from above penetrated the roof of the test vehicle but fragmented on impact. All debris missed the target entirely.

Body work presents good cover against U.S.-type slugs and sabots. Brennekes, however, went right through and struck the target intact (top left and center). As shown in bottom left photo, standard and sabot slugs tended to fragment during passage through bodywork. Again that nagging doubt about wound incapacitation.

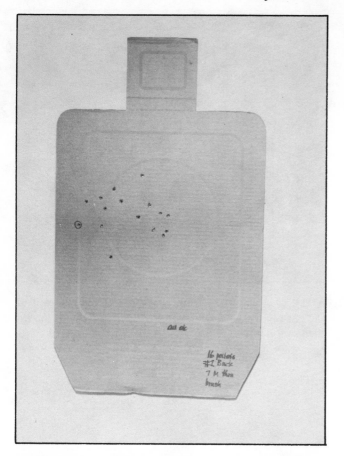

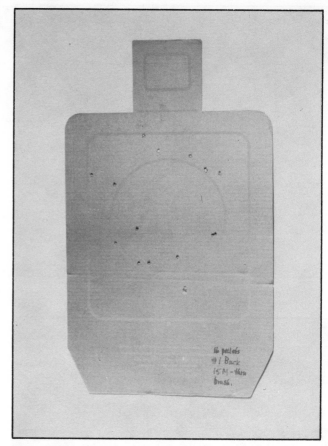

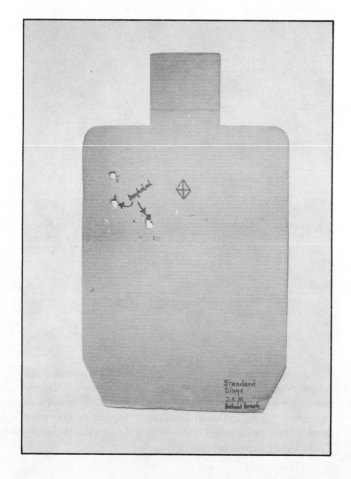

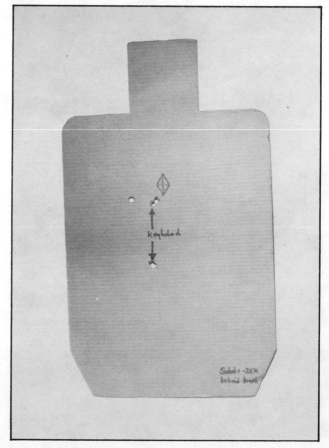

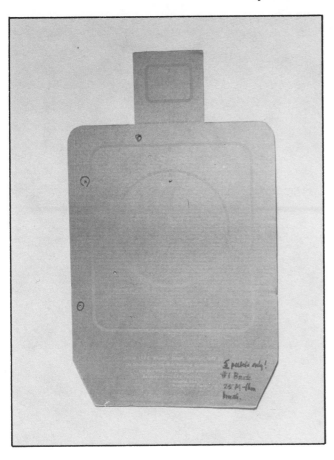

In this series of photos, left, buckshot was fired through light vegetation. At 7 meters (far left), the pattern is not much affected. The target would have been satisfactorily struck. At 15 meters (center), the pattern is still adequate, but beginning to disintegrate rapidly. At 25 meters (left), the target sustained only peripheral hits. SGN maximum effective range in light vegetation is thus about 15 meters, perhaps a meter or two beyond.

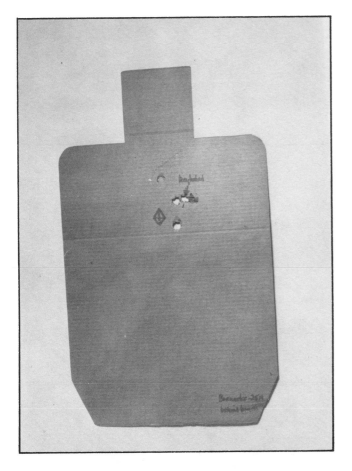

All three types of SGN slugs, the standard U.S. type (far left), sabots (center), and the Brenneke (left) demonstrated some interesting characteristics when fired through light vegetation. With the target placed directly behind the cover, all exhibited considerable keyholing and defection at 25 meters, with two out of the three shots fired. At ranges past 25 meters, those that tumbled would not have even hit the target.

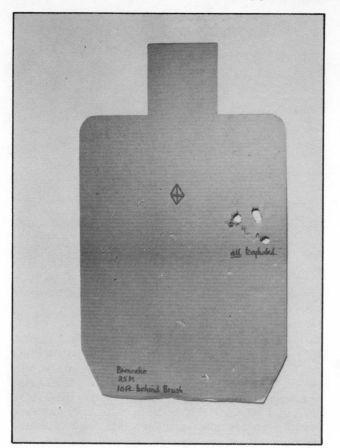

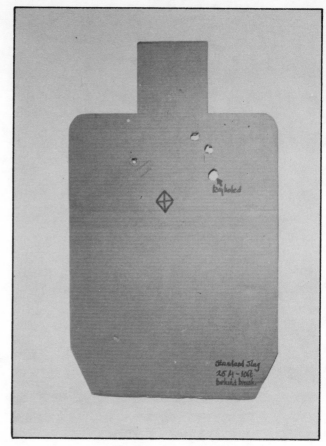

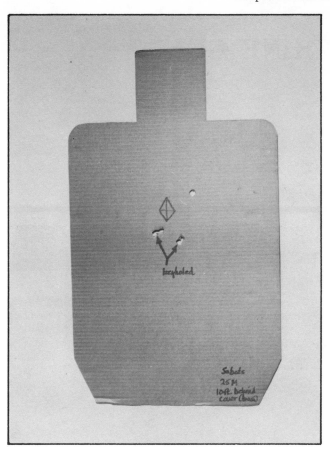

When the target was located some distance past the light vegetation, deflection and tumbling caused erratic results: Brenneke (far left), standard slugs (center), and sabots (left). As with situations where the target is located just behind the vegetation, ranges must be kept under 25 meters for reasonable efficiency.

PART B: THE SUBMACHINE GUN

SMGs are capable of utilizing a wide range of ammunition that will fulfill virtually any requirement: 9mm parabellum (top), .45 ACP (bottom).

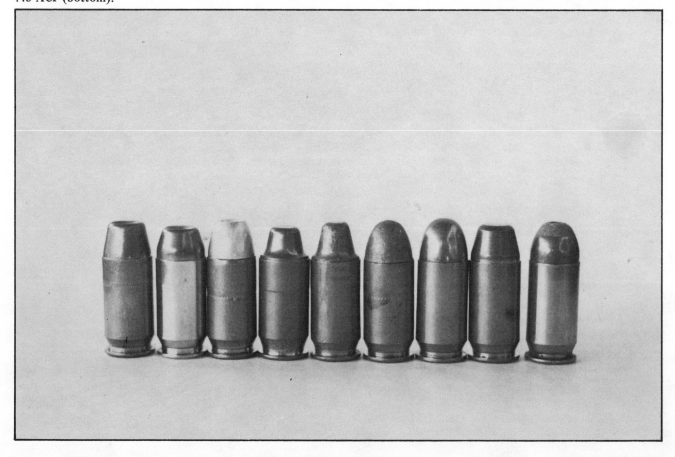

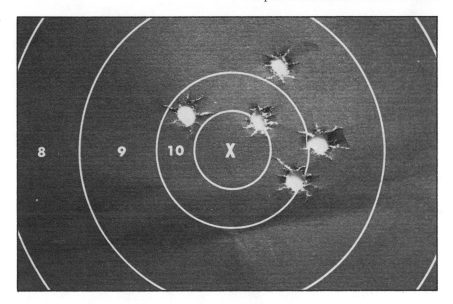

Although they fire pistol cartridges, SMGs produce significantly better performance, perhaps due to increased muzzle velocities. Shown here is a 50-meter group shot with a Sterling L-2A3 9mm SMG.

Automobile windshields present no problem to the SMG. Shown here (center and left) is a 3-shot burst from 10 meters. All bullets struck target point-on with little deflection.

Shallow angle shots across the hood from the front of the car caused ricochets into the windshield (above). The 9mm bullets penetrated the windshield and struck the target (bottom two photos) but the .45 ACP did not, instead deflecting straight up into the sky.

Not surprisingly, SMG bullets pass cleanly through side glass.

SMG fire burst through the driver's inner side-door panel. There is no doubt about serious or incapacitating wounds being produced.

Bodywork presents little cover against either 9mm or .45 ACP SMGs. This target was placed in the backseat of the car, behind the panel.

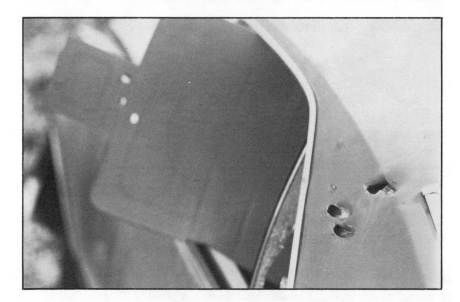

The slightly better penetration tendency of 9mm parabellum can be both an asset or a liability. The top photo shows shots from a 9mm parabellum SMG (UZI). Photo, right, shows a .45 ACP SMG (M-1928 TSMG). Useful employment can be made of ricochets if careful analysis of various bullets is made in advance.

The magazine of a STEN Mk II 9mm SMG was emptied through this windshield. Draw your own conclusions!

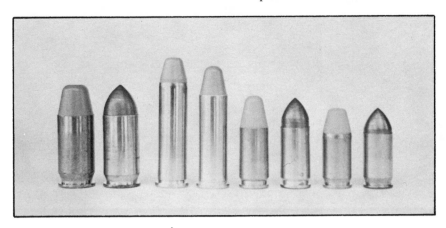

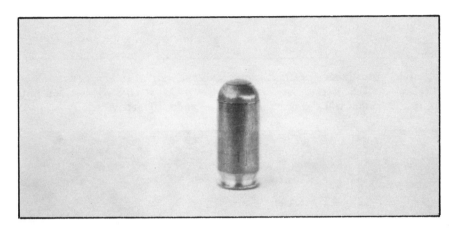

Specialized ammunition of the armor-piercing, explosive, and Glaser type significantly increase the SMGs effectiveness for law enforcement use: armor-piercing (top), Glaser Safety Slug (center), and explosive (left).

For use against automobiles, military or KTW AP ammo enhances SMG effective range. Shown in photo is a recovered .45 ACP KTW (left), a Czech 9mm AP (center), an American Ballistics Co. 9mm parabellum AP (right).

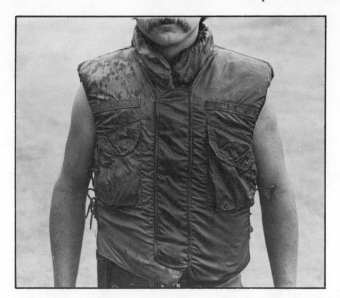 

Light body armor, such as this GI flak jacket, offers virtually no protection against SMG fire, even at 100-meter range. Most common light armor are the U.S. Army (left) and the U.S. Marine Corps ceramic plate (right) vests. SGN slugs penetrate easily at all ranges that one can actually hit the vest. Buckshot will also penetrate at ranges under 10 meters.

A KTW 9mm parabellum bullet fired from 7 meters lodged in a Second Change K-30 insert. Closer range might have allowed slightly better penetration, but this is unlikely.

Civilian "soft" armor will stop buckshot and slugs in Threat Levels II and above-rated vests. Extreme close range buckshot and any slug impacts can cause unpleasant side effects such as broken ribs and possible internal injuries. SMG fire will usually penetrate Threat Level II vests with European 9mm parabellum or American AP loadings.

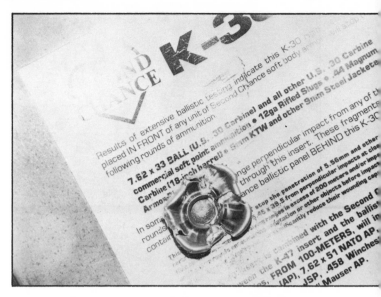

These 9mm hollow-point bullets fired from a S&W M-76 SMG were recovered from a Second Chance Model Y vest. Of the eighteen layers of Kevlar used in this particular example of soft body armor, bullets only penetrated to the seventh layer, where they lodged.

A 9mm parabellum ball expended against a K-30 insert. Virtually no SMG or SGN fire will defeat such armor, even at point-blank range.

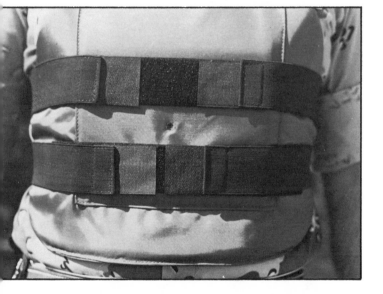

Typical "semi-hard" armor, with insert armor plate clearly visible. Bullet hole is from 9mm parabellum ball fired from S&W M-76 SMG during SWAT magazine body armor tests. The author sustained hits from various weapons and ammunition while wearing soft and hard armor.

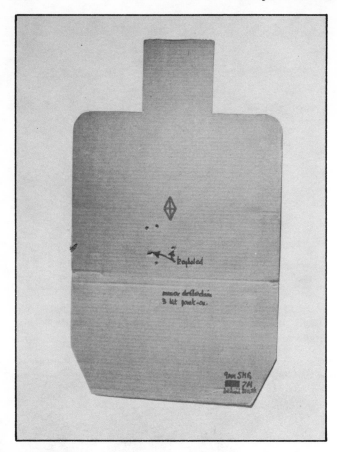
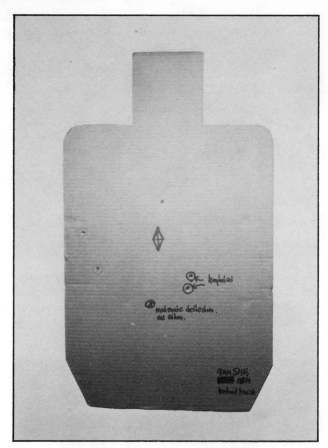
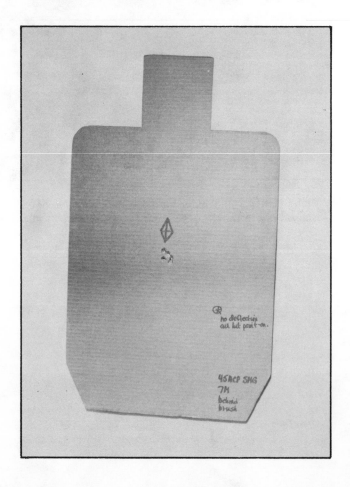
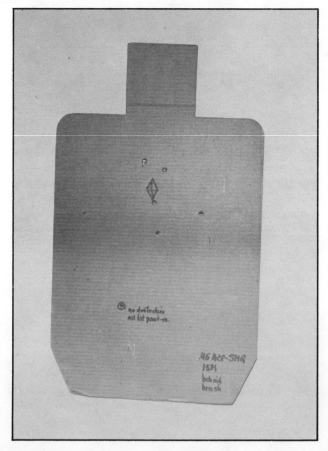

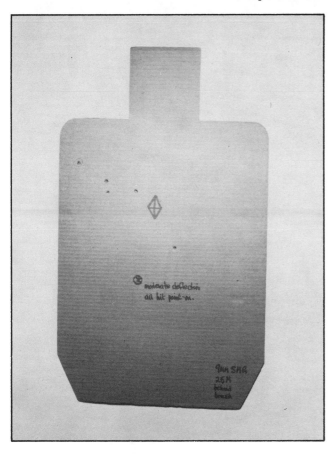

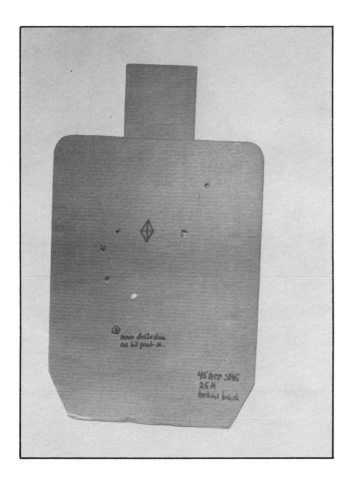

This series of photos show that light vegetation proved to be of no real threat to SMG performance. The general trend is that the .45 ACP SMG will drive right through such cover and impart incapacitating and/or lethal wounds. The 9mm parabellum tends to defect a bit more, but still performs adequately: a 9mm parabellum SMG (UZI) fired at a target placed directly behind light vegetation from 7 meters (top, far left); from 15 meters (top, center); from 25 meters (top, right); a .45 ACP SMG (M-3A1) fired at a target directly behind light vegetation (bottom, far left); from 15 meters (bottom, center); from 25 meters (left). No .45 slugs were deflected enough to miss vital areas of the target. This kind of performance is difficult to surpass—100 percent effectiveness! The 9mm parabellum was only slightly less satisfactory.

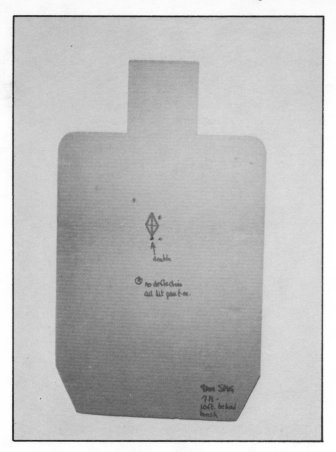

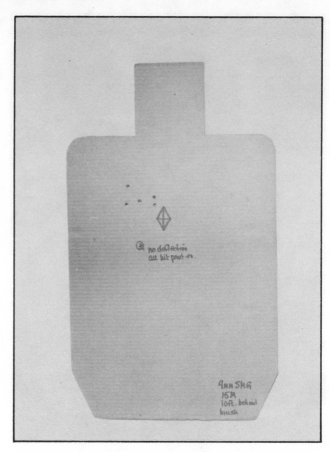

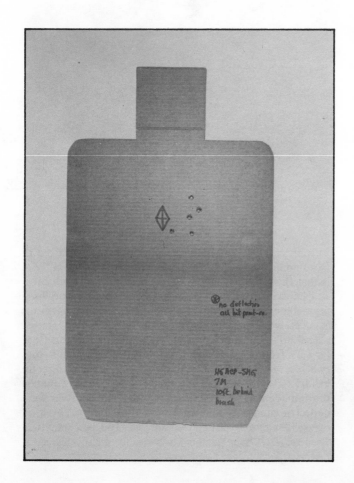

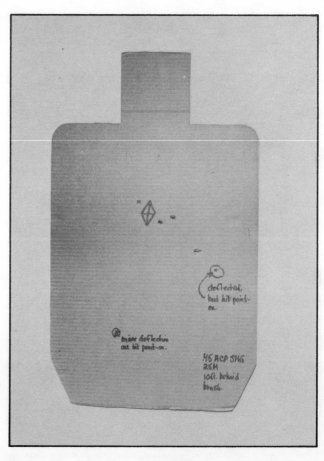

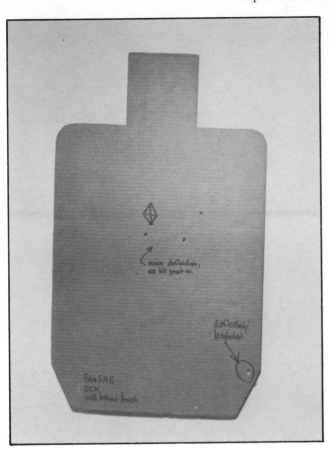

This photo series shows that when a target is some distance beyond vegetation, as in jungle or other wooded areas, the good performance of the SMG continues, with .45 ACP chambering again proving to be somewhat better than 9mm parabellum: 9mm parabellum SMG (S&W M-76) from 7 meters, target placed 10 feet beyond vegetation (top, far left); fired from 15 meters (top, center); from 25 meters (top, left). Note serious deflection; .45 ACP SMG (M-1A1 Thompson) from 7 meters, target located 10 feet past vegetation (bottom, far left); from 25 meters (bottom, left). Even a partially deflected bullet still hit the vital area. Effective range for SMG under these conditions exceeds 40 meters for .45 ACP chamberings and 30 meters for the 9mm parabellum.

4. COMPETITION AND TRAINING

As the interest in combat skills with all small arms has risen in both intensity and proliferation, so has the "other side of the coin," the pursuit of competitive activities that are merely ego-satisfying, rather than constructive in their purpose. Freud and many of his psychiatrist followers have for many years opined that guns have an almost sexual power over men. While I do not know that I agree with this, at least in its entirety, I *do* know that I have seen a number of supposedly "practical" types of competition that completely contradict reality, common sense, and tactics for no other reason than to satisfy the egos of the winning participants. Yes, perhaps they were just having fun. Who knows? But one thing is glaringly apparent. What they are doing is in no way practical and is, in fact, often actually detrimental to survival. So, inasmuch as I am interested in competition solely as a training instrument, I have no choice but to make this observation in print.

I also have some suspicions as to why this has occurred, the most prominent of which is that there are simply not enough people who design courses of fire for competitive purposes who actually know anything about tactics. Thus, in their effort to be different, they sacrifice the heart of the competitive concept. Each match need not be different just for the sake of being different. It would be far better to shoot the same course of fire over and over again, so long as that course was based upon the idea of building solid, basic shooting skills.

This problem began with handguns, but has spread to shoulder arms too, since the "practical" concept of competition possesses considerable romantic appeal to the novice. The International Practical Shooting Confederation, for example, has dominated this field for some years and is perhaps the most guilty of the deficiencies previously outlined. Speaking as a weapons and tactics professional, most of what I see at IPSC handgun matches scares me to death. What I see at IPSC rifle—and now, SGN/SMG—competitions demonstrates conclusively to me that on a collective basis, IPSC shooters are under the impression that rifles, shotguns and SMGs are just "big pistols," and competitive courses of fire dealing with them should therefore be the same with the ranges increased accordingly. Nothing could be further from the truth. Rifles, shotguns and SMGs are *not* just "big pistols," and require courses of fire all their own.

The most glaring factor to emerge from all this is that the administrative body of IPSC, long dominated by the "gamesman" category of shooter, fails to even recognize the basic purposes for which these various arms are intended. They attack with pistols, a weapon useful only as a reactive, defensive arm. Then they treat the rifle, SGN and SMG as if they were just larger versions of the same weapon. This is dangerous if realistic combat skill-building is the goal. One must understand these weapons and carefully construct courses of fire for them that promote the growth of expertise, point out the strengths and limitations of these arms in realistic situations, and indoctrinate the shooter with an understanding of tactics. In this, IPSC has failed miserably.

So then, if we can't get proper training from IPSC, who can we get it from? Get it from yourselves, that's where. Set up your own training and competitive programs based upon a serious evaluation and discussion of your needs. Remember to build basic skills—logical ones, not those gleaned from the movies or television—for these will, more than anything else, save your life when the heat is on. Don't forget to inject tactics into the competition, for regardless of how refined and efficient your techniques become, you will die in a fight unless you channel those abilities into the proper tactical avenues. If you don't *see* a target, you can't shoot it. If you fail to properly *identify* the enemy, you may shoot the wrong person and also get shot by the enemy.

Above all, keep your training program simple in scope, for rarely in real life are things so complex that they require infinitely detailed simulations in order to prepare for a confrontation.

There is no secret to effective training. All one must do in order to create training and competitive programs that maximize skill development is to remember at all times the dangers I have discussed and avoid falling prey to them. This might mean keeping a tight rein on who runs matches or teaches classes. It might also mean going to a school where the instructors are qualified to teach what you came to learn. Be careful here, too, for few can legitimately claim such expertise.

You can also read and learn from history. There have been thousands of pages written on what happens in various kinds of combat with all kinds of weapons. Dismiss what doesn't apply to you and digest fully what does. I think you will find that the creation of an intelligent program is less difficult than you think.

On the following pages are some tips that will assist you in developing an effective training program. Again, remember to simplify your program to realize the highest possible degree of skill development.

BASIC DRILL: SHOTGUN

1. Buckshot.

    a. 15 meters, single shot, perform 5 times on single target, 2.0 seconds.
    b. 10 meters, single shot, perform 5 times on single target, 1.5 seconds.
    c. 7 meters, single shot, perform 5 times on single target, 1.0 seconds.
    d. 3 meters, single shot, underarm only, perform 5 times on single target, .7 seconds.

    e. Pivots and turns, perform 5 times on single target, 7 meters.
        1. 90 degrees R—1.5 seconds
        2. 90 degrees L—1.5 seconds.
        3. 180 degrees—2.0 seconds.

    f. Multiple targets, one shot on each from 5 meters, perform twice, underarm only.
        1. 2 targets, 1 meter apart, center to center, 1.5 seconds.
        2. 3 targets, 1 meter apart, center to center, 2.0 seconds.
        3. 4 targets, 1 meter apart, center to center, 2.5 seconds.

    g. Malfunction drills, perform 5 times each, followed by single shot on single target from 5 meters, underarm only.
        1. Failure to eject, time limit: 3.0 seconds.
        2. Shell jumps shell stop, time limit: 6.0 seconds.

    h. Reloads, from empty gun, 7 meters, underarm assault only, load and fire single shot at single target, perform 5 times. Time limit: 3.0 seconds.

2. Slugs.

    a. 50 meters, single shot on single target, 3.5 seconds each, perform 10 times.
    b. Head shots, 7 meters, single shot only, perform 5 times. Time limit: 2.0 seconds each.

Possible score: 415 points.

Penalties: Overtime shot or drill completion—minus 5 points.

Knockdown steel silhouettes cut to the external dimensions of the Taylor Combat Target are recommended. If paper targets are used, score major caliber. If majority of shot pattern is inside vital area, score 5 points. If not, score 3 points. If steel silhouette is struck hard enough to be knocked down, score 5 points per. If not, score as miss. Slugs should be scored 5/3 for body and head alike.

ADVANCED DRILL: SHOTGUN

1. Buckshot, all single shots.

    a.   15 meters, single shot, perform 5 times on single target, 1.2 seconds.
    b.   10 meters, single shot, perform 5 times on single target, 1.0 seconds.
    c.   7 meters, single shot, perform 5 times on single target, .7 seconds.
    d.   3 meters, single shot, perform 5 times on single target, .5 seconds.

    e.   Pivots and turns, perform five times each on single target, 7 meters, single shots.
         1.   90 degrees R—1.2 seconds.
         2.   90 degrees L—1.2 seconds.
         3.   180 degrees—1.8 seconds

    f.   Multiple targets, one shot on each from 5 meters, perform twice, underarm only.
         1.   2 targets, 1 meter apart, center to center, 1.0 seconds.
         2.   3 targets, 1 meter apart, center to center, 1.3 seconds.
         3.   4 targets, 1 meter apart, center to center, 1.7 seconds.

    g.   Malfunction drills, perform five times each, followed by single shot on single target from 5 meters, underarm only.
         1.   Failure to eject, time limit: 2.0 seconds.
         2.   Shell jumps shell stop, time limit: 4.5 seconds.

    h.   Reloads, from empty gun, 7 meters, underarm assault only, load and fire single shot at single target, perform five times. Time limit: 2.0 seconds.

2. Slugs.

    a.   50 meters, single shot on single target, 2.0 seconds each. Perform 10 times.
    b.   Head shots, 7 meters, single shot only, perform five times. Time limit: 1.5 seconds each.

Targets, scoring, and penalties same as with basic SGN drill.

BASIC DRILL: SUBMACHINE GUN

1. 50 meters, 2 shots on single target, 4.0 seconds each, perform twice.

2. 25 meters, 2 shots on single target, 2.5 seconds each, perform twice.

3. 15 meters, 2 shots on single target, 1.7 seconds each, perform twice.

4. 10 meters, 2 shots on single target, 1.5 seconds each, perform twice.

5. 7 meters, 2 shots underarm assault only, 1.0 seconds each, perform twice.

6. 3 meters, 2 shots underarm assault only, .7 seconds each, perform twice.

7. Multiple targets, 2 shots on each, 5 meters, perform twice, underarm only.
   a. 2 targets, 1 meter apart, center to center, 1.5 seconds.
   b. 3 targets, 1 meter apart, center to center, 2.0 seconds.
   c. 4 targets, 1 meter apart, center to center, 2.5 seconds.

8. Pivots and turns, 7 meters, 2 shots fired on each string from underarm assault only, 5 repetitions each.
   a. 90 degrees R−1.5 seconds.
   b. 90 degrees L−1.5 seconds.
   c. 180 degrees−2.0 seconds.

9. Malfunction drills, followed by 2-shot burst on single target from underarm assault only.

10. Reloads, NO SHOOT, perform 5 times, time limit: 5.0 seconds.

11. Head shots, 7 meters, perform 5 times, single shots, time limit: 2.0 seconds each.

Possible score: 535 points.

Penalties: Minus 5 points for any overtime shot or drill completion.
             Minus 5 points for any complete miss.

Knockdown steel silhouettes cut to the external dimensions of the Taylor Combat Target are recommended. If paper targets are used, score major caliber for .45 ACP and minor caliber for 9mm parabellum. If steel targets are used, score 5 points for each target knocked down. If target fails to fall, score as miss.

Targets should be set to fall with solid hit from 9mm parabellum ball ammunition fired from a handgun at 50 meters.

ADVANCED DRILL: SUBMACHINE GUN

1. 50 meters, 2 shots on single target, 3.0 seconds, perform twice.

2. 25 meters, 2 shots on single target, 2.0 seconds, perform twice.

3. 15 meters, 2 shots on single target, 1.2 seconds, perform twice.

4. 10 meters, 2 shots on single target, 1.0 seconds, perform twice.

5. 7 meters, 2 shots on single target, .7 seconds, perform twice.

6. 3 meters, 2 shots on single target, .5 seconds, perform twice.

7. Multiple targets, 2 shots on each, 5 meters, perform twice, underarm assault only.
    a. 2 targets, 1 meter apart, center to center, 1.0 seconds.
    b. 3 targets, 1 meter apart, center to center, 1.3 seconds.
    c. 4 targets, 1 meter apart, center to center, 1.7 seconds.

8. Pivots and turns, 7 meters, 2 shots fired on each string from underarm assault only, 5 repetitions each.
    a. 90 degrees R—1.2 seconds.
    b. 90 degrees L—1.2 seconds.
    c. 180 degrees—1.8 seconds.

9. Malfunction drills, followed by 2-shot burst on single target from underarm assault only.
    a. Position 1—1.5 seconds, perform twice.
    b. Position 2—1.5 seconds, perform twice.
    c. Position 3—4.5 seconds, perform twice.

10. Reloads, NO SHOOT, perform 5 times, time limit: 4.5 seconds each.

11. Head shots, 7 meters, perform 5 times, single shots, time limit: 1.5 seconds each.

12. Head shots, 10 meters, perform 5 times, single shots, time limit: 1.7 seconds each.

Possible score: 560 points.

Penalties: Minus 5 points for any overtime shot or drill completion.
             Minus 5 points for any complete miss.

"Basic drill," that portion of a training regimen that develops fundamental skills, should be the heart of any training program with the SGN or SMG. Here the author works out on multiple targets with S&W M-76 SMG. SWAT magazine staffer Ken Hackathorn demonstrates (previous page) an excellent Taylor offhand with his Benelli M-121 SGN. Note minimal muzzle rise with this particular technique.

Indoor reaction ranges bring an added element of stress and flexibility into the training curriculum as long as the scenarios used reflect both realism and practicality. Loading up at a "haunted house" with scattered targets without regard to building tactical skills is nothing more than a "circus" which, while it might be fun, is of no value in building survival skills.

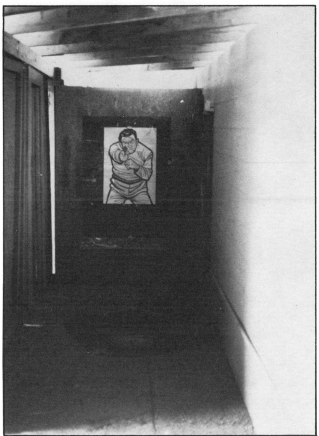

Example of well-placed target. The student must enter the dwelling and acclimate his eyes to indoor light conditions, then begin clearing the building. The target is located approximately 15 meters at the end of a hallway and is actuated via a pull cable hidden from the shooter.

Closets also present a hazard because of their closeness. They, too, should be a part of any indoor reaction range.

Corners present a serious hazard to anyone forced to engage an enemy indoors. For this reason, any indoor scenario should include proper negotiation of them.

Outdoor reaction ranges, such as the example shown, are of limited value unless the terrain in which they are located is in keeping with the missions for which the weapons being used are intended. This particular range is of very limited usefulness because it ignores this premise.

Targets used on outdoor ranges should be placed the way real enemies would place themselves. This means partial conceal-ment at a minimum. You can't stay alive, regardless of how good a shot you become, if you don't train yourself to detect targets.

Instead of confining participants in reaction drills to a road, a path through vegetation is a better idea. This allows the participant to use principles of cover and concealment, whereas the road forces him to stay mostly in the open.

While useful if properly utilized, complex "props" (above and below) are largely unnecessary to administer good training programs.

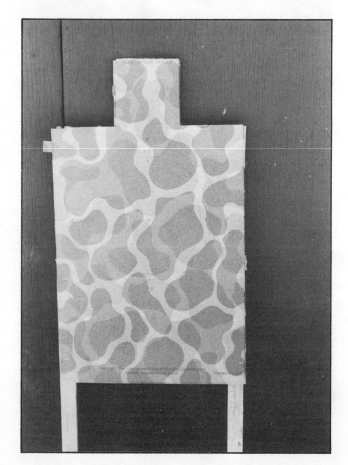

Type of targets used in any kind of training is one of the most critical, yet overlooked facets of instruction. The general rule is that the more different kinds of targets you use, the better off your student will be in the real world. A selection of targets is shown on this and the opposite page.

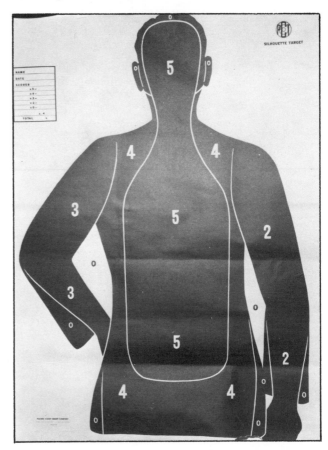

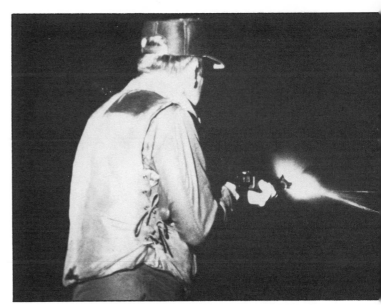

Any properly designed training program will include night firing. A great deal of action with both the SGN and SMG takes place during dim light conditions and it is important that the student be able to handle himself under such conditions.

5. SHOTGUN AND SUBMACHINE GUN TACTICS

Modern combat small arms skills have reached a level of efficiency that far surpasses those of the past. Indeed there are more excellent shots in existence today than ever before—as far as technique is concerned, that is. No, I'm not knocking these men, far from it, because they are very good shots and many of them have critically influenced the very evolution of weaponcraft with their competitive discoveries of winning technique. Yet, most of them lack the most important element of survival skill: an understanding of tactics.

If one places himself at the wrong place at the wrong time, his survival against an armed opponent hinges upon forces that have nothing whatsoever to do with his own skill. A great shot cannot hit what he does not see. The fastest shot in the world will still go to prison if he fails to properly identify his target and shoots the wrong person.

It is no secret that under stress people revert to a completely nonintellectual condition known as the conditioned response. This, in fact, is the entire reason for training oneself—to program the proper conditioned responses to the widest possible variety of stimuli. Almost never is there enough time to understand intellectually the events that take place in a fire fight because there just isn't time.

We have already established that the handgun is by nature a defensive weapon. More powerful weapons like the rifle, shotgun and submachine gun perform offensive duties in a fashion far superior to the handgun because the handgun is more difficult to use well, less powerful and lacks flexibility.

At the outset we can say that the marksmanship problem experienced during a typical fire fight is relatively easy, and this brings us to the point: brilliant marksmanship isn't necessarily a guarantee of survival in a combat situation. But common sense goes a long, long way to enhance the odds! In what follows we will explore this largely misunderstood but absolutely critical field in detail, with the hope of providing the reader with maximum understanding.

Tactics are not a mixture of voodoo and alchemy, as entirely too many people are under the impression they are. Tactics merely reflect a crystallization of the various factors of common sense that apply under any given set of circumstances. In-depth programming of the subconscious mind to react with proper responses to those circumstances is what training is all about.

First I must point out that what will be disclosed herein is not a *guarantee* of survival— only of a reduction of danger to controllable levels. It is entirely possible that you can find yourself in an unsurvivable scenario, although statistically it isn't very likely, and no amount of tactical savvy and/or marksmanship skill can save you.

There are six basic parameters of tactical awareness. Adherence to them will do much to minimize your danger and will make your efforts more effective.

Number 1 is to *use your eyes and ears.* Your brain is like the headquarters of a large organization—it must have data to recognize a problem, consider possible courses of action, select a course of action from those possibilities, and, finally, embark upon that course of action. Your senses of sight and sound can provide you with invaluable data and, indeed, do just that all the time. We simply ignore that data—i.e., we look but we don't see and we listen but we don't hear. We do this especially in an urban environment because we tend to shut out that environment as being basically artificial, which it is. This doesn't mean that elements existing within that environment cannot prey upon us.

When was the last time you paid attention to who might be sitting in that car that just pulled up alongside you at the stoplight? Or, how about the vehicle that appears in your rearview mirror? With urban terrorism on the increase, one cannot afford to ignore what goes on around him for it may indeed prove dangerous, even fatal. The purpose of the firearm is to provide you with the means to control your immediate surroundings. If you are not paying attention to those surroundings, how can you do this?

Along these same lines, the presence of an extraordinary noise might well be an indication of imminent action. The scraping of fabric along a wall in that darkened living room, or the sound of someone dragging his feet along the ground or floor. People tend to do these things when under stress and if you have trained yourself to listen for and recognize them, the element of surprise is on your side instead of your attacker's.

When searching an urban structure or an outside area, be systematic and thorough in your efforts. Don't let your eyes wander about, drawn by prominent objects or bright colors. Instead, search in and out on an axis by changing the focus of your eyes along a straight line. This way even someone not directly in your line of vision will be seen.

Number 2 is *never turn your back on anything you haven't checked out first.* Put another way—don't assume anything, check it out. People can hide in the wildest places if they feel they must, ranging from behind clothes and shoe racks in closets, to suspending themselves by their hands and feet from the ceiling of a small room, to crouching on the shelf of a closet.

I know of an instance where a burglar was observed entering an urban dwelling by neighbors. The police were notified and a number of officers responded by encircling the house. In the time it took the police to arrive, several minutes, the neighbors stood watch on the place, ensuring that the suspect had not left the premises.

When the police searched the house they were unable to find the suspect, even after hours of effort, and finally withdrew under the impression that the suspect had somehow escaped. Fortunately, as the last few officers were making their way back from the interior of the house to the front door, a tiny closet opened, bumping one of the officers as he passed and, from that closet, emerged the burglar! The door to the closet had been opened several times by searching officers, but none had bothered to investigate further, thinking that the contents of the closet itself precluded the presence of anything as large as a human being. The suspect had been behind some hanging clothes and had remained undetected for almost four hours!

When you are entering a room, make certain that you see all of it—walls and ceiling—before you turn your attention to other things. Remember that you cannot shoot what you cannot see. But, on the other hand, what you cannot see *can* attack you with considerable likelihood of success.

The *third rule* of tactics is to *stay away from corners.* Constantly violated by TV and cinema stars, this principle has assumed the proportion in my mind of being one of the most serious facets of combat tactics, perhaps because of my own experience with it.

I once observed a young infantry officer in Vietnam who had neutralized, we all thought, a North Vietnamese machine gun bunker with a hand grenade after carefully maneuvering to within close range of it. It was common knowledge that most MG teams were made up of two, sometimes even three men, and, from the volume of fire that had been sustained from the bunker, it was almost certain that there were a number of NVA inside.

Carefully moving along the left side of the now-gutted bunker, the officer, armed with a Thompson submachine gun, made his was to the rear of the structure where the only means of ingress and egress, the rear door, was located.

It was an easy "mop up" maneuver, for there was little danger that anyone inside the bunker had survived the explosion of the hand grenade tossed inside, so the officer was casual in his approach to the left rear corner of the bunker and stepped around it directly at its apex instead of placing himself well away from it and stepping out with his weapon ready.

At the exact instant the officer negotiated the

corner, an NVA soldier armed with an AK-47 assault rifle emerged from the bunker. The U.S. officer sustained three hits from the AK and was able only through miraculous luck to kill the NVA with his submachine gun before falling to the ground unconscious from his wounds.

This particular incident well illustrates how people get hurt unnecessarily by failing to adhere to common sense tactical principles. The officer should have kept well back from the corner and when he finally exposed himself across the angle projecting from it, he should have done so with his weapon at the ready. Had this been accomplished, the NVA would have been instantly neutralized with no injury whatsoever to the U.S. officer.

This incident drove home to me the importance of this tactic—*because I was that young officer!*

Rule *number 4* is to *maximize the distance between you and any potential danger area as much as possible.* The closer you are to an attacker, the easier it is for him to get you. Close ranges reduce time frames, giving you less time to react and respond. If the attacker has an edged or blunt weapon, he can effectively use it against you. If he has a gun, it will be easier for him to place a hit on you. A man across the room with a knife is no problem to someone armed with a firearm. On the other hand, if his location is within a step or two he can bring his weapon into action against you with some possibility of success.

The way to move about when engaged in combative activities is the Taylor Ready position as illustrated on page 114. Remember not to drag your feet or place your back against a wall as you move—that potential enemy has ears too—and to use your eyes and ears at all times. Don't forget to keep your finger off the trigger while "hunting." The gun can be brought to a firing condition as quickly as it can be brought to bear on the target.

*Keep your balance—rule number 5—*don't be caught leaping around. You cannot bring effective fire upon an assailant without keeping yourself under control. Move briskly across danger areas but do not run unless already under fire. Move in a sideways shuffle, maintaining balance as you go. Keep the direction of your weapon generally in line with where you are looking. Do not cross your legs.

Don't forget the possibility of encountering unarmed or non-hostile personnel exists. For this reason you must never fail to positively identify your target before taking it under fire. This is easily done in the time that it takes to move your weapon from the "ready" to the "point" position. It should also be noted that the law in most states dictates that you must be in fear of your life or immense bodily harm before responding with deadly force. A man armed with a knife or club standing across the room may not legally constitute a deadly threat.

The last basic rule is to *watch your front sight.* This one isn't as easy to *do* as to say because your eyes will be telling your brain to look at the source of excitement. You must fight that tendency and, no matter what, keep your eye focused on the front sight. If you do, you'll hit him, if not—cast your fate to the winds, maybe yes, maybe no. The point is that you do not have to take unnecessary risks.

Other procedures of interest include doorway entry and clearance, window clearance, malfunctions, and reloading.

Doorway entry is a simple procedure that requires only a bit of calm and method to execute. First determine which way the door opens—in or out—and station yourself on the side *opposite* the hinges. If the door opens inward, the hinges will most likely be inside and thus invisible to anyone located outside. If it opens outward, they normally show. In those cases where one may not be able to see the direction which the door travels toward, the doorknob is a good indicator.

The reason for placing yourself on the side opposite the hinges is to allow maximum visibility into the interior as soon as the door begins to swing open, thus providing you with the earliest possible advantage of detection and subsequent reaction.

Once the door is moving, ensure that it moves all the way to the wall, in order to disclose the presence of anyone behind it. If someone is indeed there, he must either institute action prematurely (you're still outside, remember?), get his face bashed in by the door, or come out. In any case, the advantage remains with you. Then get inside and to either side to avoid silhouetting yourself (this holds true for windows as well).

Resist the impulse to peek through windows. Treat them the same way as you would corners, i.e., keep a maximum distance from them at all times.

Unless the need to reload comes in the middle

of an exchange of fire, don't drop your expended magazines on the ground. If you must reload, be sure to take cover unless you are in the middle of an exchange.

You should not shoot your weapon empty unless the tactical situation demands it. It will take longer to reload. Reload the SMG after no more than five short bursts. The SGN must be continually reloaded as its ammunition is fired in order to maintain sufficient quantities in the magazine.

If you have a malfunction, take cover to clear it. Don't stand there like a target and "fiddle around" with your gun. Rehearse malfunction-clearance drills until you are completely familiar with them. They could save your life.

Remember that just because you are inside, that doesn't mean that your opponent is also inside. He may have slipped outside as you entered or moved about after entry and be lurking nearby, perhaps waiting for the opportunity to get a chance at you through a window or plate glass door. When we are inside we tend to pay no attention to this possibility although failure to do so could result in serious injury or even death.

If you do find yourself in a fight, handle it as swiftly and aggressively as possible. If police officers arrive on the scene, ensure that your weapon is either placed on the ground or held in a nonthreatening posture. Police officers are human beings too, just like you . . . and they have no idea what is going on when they arrive on the scene—only that there as been a call of "shots fired." They have no way of knowing what has happened and, as a result, they tend to be very careful about endangering themselves, just as you would be. Don't think that they will somehow automatically know that you were the intended victim, because there is no way that they could. Be calm and

remember that, although the police are not "out to get you," statements you make become a matter of official record. Tell the truth correctly and without alteration and do not feel that you are incriminating yourself by having your attorney on the scene when you tell your story. Some authorities go so far as to advise that you call your attorney first—then call the police. I personally opine that such a decision rests with the participants involved in the altercation, based upon the particular circumstances.

Other authorities state that you should automatically prepare yourself to "be sued" by the family of the person you have shot in your own defense. With this I categorically disagree. While it is possible that civil litigation against you could be instituted by the survivors or relatives of the perpetrator of the attack, remember that *you* were the intended victim and that legal justification of the use of deadly force in your own defense is a more than adequate defense against civil suit for that act. More often than not, this fact will prevent civil suit against you, if anything, because those who are considering the suit will be advised of this by their own attorney. Besides, as an associate of mine recently said, "If it boils right down to it . . . I'd rather be tried by twelve than carried by six." The extreme simplicity and basic honesty of this statement needs no further explanation.

These tips will do much to reduce your personal risk when faced with a tactical situation. Careful adherence to them will make your job much easier, thus freeing your mind to ponder things. "Murphy's Law" and the "KISS" principle both apply to all forms of serious combat and anything you can do to alleviate potential problems before they occur will increase your own survival potential.

Stay away from corners! You cannot respond to an attack when caught in such a position because you will be detected only at extremely close range, giving you little time to react and respond. Also, if the wall is of hard substance, bullets or shot can ricochet into you. Stay well back from the corner itself to avoid this.

Always ascertain that the door through which you wish to gain entry is indeed negotiable. Do this by carefully checking it. If the door is indeed useable, station yourself on the side away from the hinges to allow maximum visibility as the door swings open under your push. Be ready for action. Once ingress has been gained, get inside and away from the entrance to avoid silhouetting yourself.

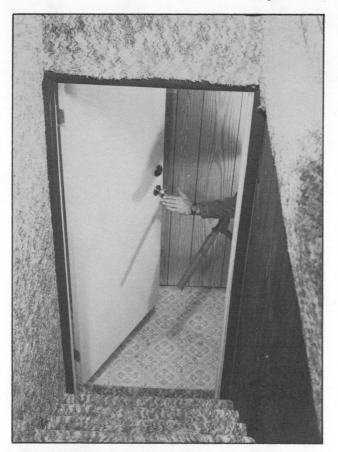

Be very careful about giving up your mobility during any tactical situation because it is your biggest asset. Any action sustained under the situation shown here will result in injury to you.

Stairways constitute an especially serious hazard because they channelize your movements and bring engagement ranges down to very close distances. Be particularly careful when negotiating such hazards, regardless of whether you are travelling up or down stairs.

Counters also should be handled with extreme caution. Be particularly alert when working with them. The WRONG way to handle a counter (above). Firer is off balance and exposed. The RIGHT way to handle a counter (below). Move in on the counter from the side and step out with your weapon ready only when you feel that you have the problem under control.

Keep your balance! You cannot return effective fire or otherwise protect yourself when caught like this. **DO NOT RUN** around the area unless you are already under fire and seeking cover or concealment.

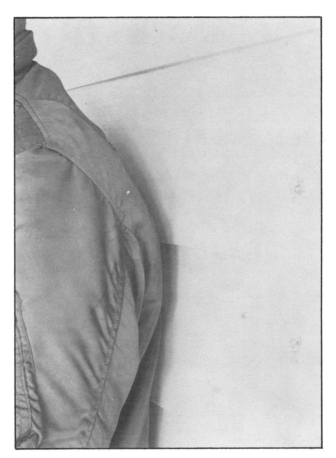

Watch out for making noise with either your back or feet. It is very typical to see people under stress plant themselves against a wall and "slither" down it. If you hear noises such as these, they cannot be friendly!

When outdoors, avoid silhouetting or "skylining" yourself— for obvious reasons!

This series of photos shows the proper way to enter a building through a sliding door.

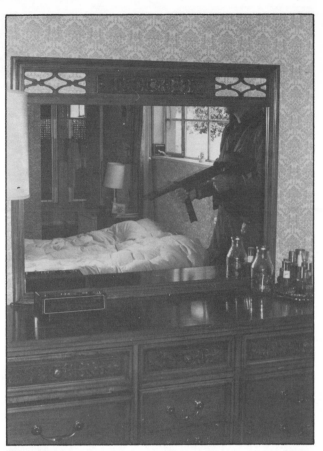

Watch out for mirrors and other reflective objects. They can give you away prematurely.

Be careful around windows. Avoid crossing them if you can. If you must, then do so quickly.

Remember that automobile side-glass is of little use as cover or concealment. This is a guaranteed invitation to get shot in the guts!

Don't lean across the hood of a car or any other hard object. Your opponent can easily ricochet bullets or buckshot into you.

Station yourself well back from your cover. This way any ricochets will pass over your head instead of into it!

Use the tires, wheels, and engine compartment as much as possible to provide you with real protection against incoming fire. This often requires cramped, uncomfortable shooting positions.

Understand that your feet and part of your legs are exposed if you stand like this. As with any other hard object, bullets can ricochet off the ground into you.

Car doors themselves provide protection only against most handgun bullets and buckshot. SMG and rifle fire, as well as some of the more potent handguns, will easily penetrate car doors.

Rocks used as cover should be handled like any other hard object. Don't get too close or ricochets can hit you. Instead, place yourself to the rear of the object to allow ricochets to pass over you.

Know the difference between "cover" and mere concealment. This bush may partially conceal you, but it certainly will not protect you.

Trees are not as bullet resistant as many believe. When you use one for cover, make sure it's as big a one as you can find!

6. TECHNIQUES FOR WEAPON EFFECTIVENESS

There have been a great many different theories over the years about how best to utilize the SMG and shotgun for "social" functions. Sadly, an analysis of them usually discloses that their actual effectiveness ranges from marginally adequate to potentially disastrous.

Even the military are of little help here, for in spite of their mammoth resources and virtually unlimited budgets, Uncle Sam has developed a rather negative attitude towards training of any kind during the last several decades, particularly when it comes to small arms.

The situation is complicated by a myriad of writers, all seeking to make their mark in the business. This is fine with me, as long as any technique they espouse is legitimate. All too often—as a rule, in fact—we find that, much to our chagrin and potential demise in some cases, these individuals really didn't know what they were talking about.

I know how this feels, for many years ago I stood in those shoes. I was a young infantry officer, fresh from Fort Benning's Infantry School and the mud of Camp Darby, the Ranger Training Center. I was eager and confident, even a bit cocky, at least according to those of my associates who have survived to reminisce about those days. I went to combat in Southeast Asia full of techniques—only to find that they were either antiquated or just plain ineffective. I watched men die and I sustained personal injury because of this.

I made a vow to myself that if I lived and made it home in one piece, I would devote myself to finding what worked in a real fight and to hell with techniques conceived and intended for use on strictly controlled firing ranges. It has taken me years of research, experimentation and comparative interviews with people whom I know to be for real to compile and test the techniques set forth in this section. After such testing, which included real action, I am satisfied that they not only work, but are the best now known to exist. I have been able to cause students to progress far more quickly and to greater heights than I had ever dreamed possible utilizing these techniques and a proper instructional format. To my knowledge, no one who has used them has to date been killed or even seriously injured, even though the occasions on which they have now been used number in the hundreds.

I think this speaks for itself and now offer them to you. Don't forget that the key to perfection lies in repetition of fundamental skills. This means practicing what you see in the accompanying photographs until you can perform them in your sleep. When you can do that, you can take a rest, but not before!

Technique is one-third of what I call the *Triad of Survival*. As the name implies, there are three parts: *Tactics, Technique* and *Mental Conditioning*. No one element is more important than the other. On the contrary, each is equally dependent upon the others and equal emphasis must thus be placed on all three. After a perusal of this section you will realize that all of the techniques set forth are designed to act in unison.

Mastery of Technique, along with Tactics and Mental Conditioning, provides one with the maximum probability of surviving a fight. The element of luck is out of our reach. Although it unquestionably has some influence upon the outcome of any situation, we cannot depend upon it to carry us through. I always plan for the worst on the premise that if things don't get to be as bad as I have planned for, I am that much further ahead. I suggest that you do the same in order to avoid what I call "terminal surprises": unforeseen contingencies that develop during the course of action and, because planning and training were administered with the idea of having "just enough" to get the job done, become insurmountable obstacles.

The techniques I have developed are the result of years of study, experimentation, research and analysis. I did not create all of them myself, nor do I wish the reader to think that I did. Some of them have existed for as many as fifty years; others much less. What I *have* done is create a viable, effective repertoire of techniques that apply equally to the combat shotgun and submachine gun in deference to the fact that they are both light shoulder arms. This means that they are not only the most effective now known to exist, but are teachable in a remarkably short period of time with a minimum of ammunition expended. Among these techniques you will find no skeet shooter's traditional shotgun methods. Nor will you find any "spray and pray" SMG procedures, for neither of these have any place in combat. What you *will* find are simple, logical movements and methods, designed to maximize your stability and balance while at the same time lowering your response times.

Whenever the weapon is fired from the shoul-

der, the sights are *always* used, with operator attention focused on the front sight in particular. Perhaps controversial among the uninformed, but no less effective, is my Underarm Assault position, which is used at ranges not to exceed ten meters. It is here that the SMG is used in the fully automatic mode, in short bursts of, preferably, two shots. I realize that the common notion is that the SMG should be fired in bursts of three to five rounds. I do not subscribe to this belief because it is based upon the premise that the SMG is used like a fire hose, which it is not. Specialized fire missions requiring area-fire aside, I feel that ammunition conservation versus hits per shots fired is the real issue. Since the third shot of any hand-held burst is always a waste anyway, why throw away ammunition?

As with any firearm, the SMG and SGN malfunction-clearance drills are just as important as any other technique. Both these weapons perform well if used properly, but both are also machines. This means that an occasional stoppage can occur. A stoppage can be cleared in seconds without the use of tools. A "jam" requires the services of an armorer or gunsmith with tools, an impossibility in the field, especially under fire. We must call things what they are to avoid confusion.

There is quite enough confusion as it is when a fight is in progress—and if you are going to survive you need to simplify, not complicate, the problem.

LOADING THE SMG

Retract actuator to cock weapon; insert magazine.

Briskly strike magazine as shown with heel of supporting hand. Weapon is now ready to fire.

LOADING THE SGN

Grasping a live shotshell between the inside edges of the index and little fingers of the weak hand, roll the shell into the ejection port as shown. Learn to do this by feel so you can keep your eyes where they can do the most good—downrange!

Briskly close the action and set the safety.

LOADING THE DOUBLE-BARRELLED SGN

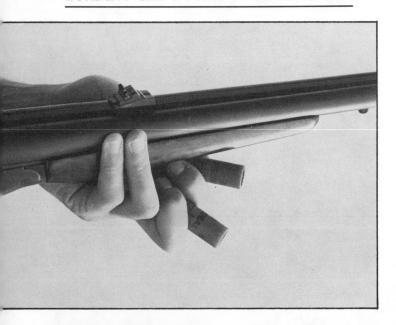

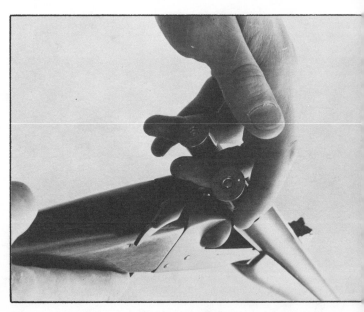

Grasping two shells between the fingers as shown, break open the weapon.

Insert both shells simultaneously into breech and let them fall home.

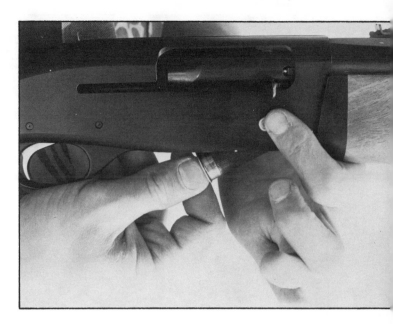

Load the magazine by feel, pressing each shell by inserting it into the loading gate and pressing it home with the pad of the weak hand thumb. Always keep the firing hand on the small of the stock as much as possible to maximize efficiency.

On many self-loaders, a button must be depressed to unlock the loading gate. While awkward, this is best accomplished using the technique demonstrated above.

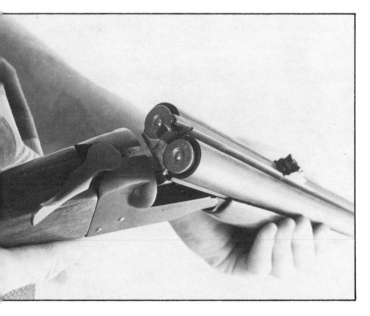

Briskly close action and set the safety. Weapon is now ready to fire.

SMG CARRYING CONDITIONS

Buttstock should always be opened and locked during periods of imminent action. Do NOT carry the SMG "cocked and locked" with the stock folded.

SGN CARRYING CONDITIONS

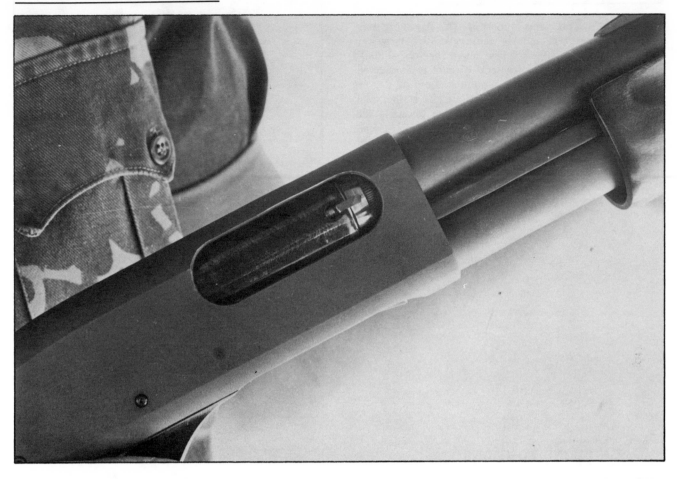

Other than with the safety on and the trigger finger resting upon it, the operator of a slide-action SGN can, by utilizing the action-release button, "crack" the action of his weapon slightly open, allowing complete safety while maximizing his reaction/response time. All he must do is briskly slam the action closed as he brings the weapon to bear on his target instead of manipulating the safety button, most of which are both too small and poorly placed.

SMG/SGN CARRYING POSITIONS

High Ready or Port Arms (opposite)—This is a short duration carry that will quickly cause excessive fatigue if attempted for too long. It is best suited for "casual alert" situations. **Taylor Ready (center)**—Weapon is shouldered, then lowered until the muzzle is pointed about 40 degrees below horizontal. Head is kept forward, as is the weak foot. This position is designed for moving toward a specific threat and giving the fastest response time. **Rhodesian Ready (far right)**—Weapon is cradled as shown with butt tucked loosely under arm. This particular position is designed for long-term carry in the field and is the least fatiguing. It is also the slowest from which to respond except at very close range, when the Taylor Underarm Assault would be used.

SMG/SGN FIRING POSITIONS

Offhand (opposite)—Weapon is shouldered, sights are used. Weak leg is forward with knee lightly flexed. Strong knee is locked, placing balance forward. Both elbows are kept DOWN at about 40 degrees below horizontal. Make certain that you hold the gun to your shoulder with the middle, third, and little fingers of your firing hand and not your supporting hand (center). This provides maximum consistency and strength which translate to control of the weapon. **WRONG position (far right)**—The SGN/SMG is NOT a rifle and is not fired like one. This firer has his balance rearward, his firing arm horizontal, and his supporting arm vertical. This won't do!

This operator is grasping the weapon around the front of the magazine housing instead of on the fore end. Control and stability are reduced.

Taylor Underarm Assault—As with the Offhand Position, the weak leg is forward, knee lightly flexed. Strong leg is trailing with knee locked. Balance is forward but not excessively so. Butt of gun is held securely by the firing arm/

INCORRECT technique places operator balance too far to the rear, reducing his mobility and weapon control. Head is too far to the rear and is not aligned with the weapon. Weak elbow is horizontal, allowing lateral drift.

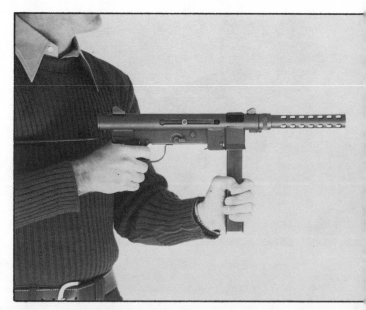

NEVER fire the SMG as shown (above and right). Always hold it by the fore end.

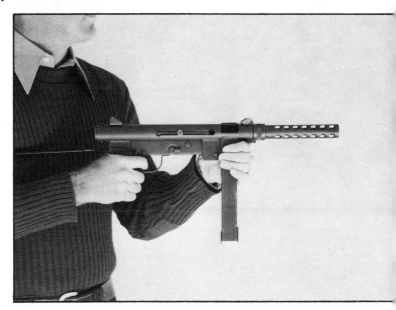

Second generation SMGs should be grasped as shown, NOT by the magazine, for maximum weapon reliability and results downrange.

elbow high under the armpit against the pectoral muscle, with the butt itself just inside the armpit. Both elbows are kept about 40 degrees below horizontal. Master eye is placed over the bore of the weapon.

NEVER employ the SMG without first deploying the buttstock. The stock should already be extended by the time actual weapon utilization occurs.

On weapons with excessively short buttstocks, a slightly more pronounced forward body attitude must be used. This is common with many folding stocked SMGs and SGNs.

On many third generation SMGs, the receiver is so short the operator is tempted to grasp something other than the fore end area. The technique shown here makes for some pretty erratic shooting!

Many shooters find that their speed and pointability is increased by placing the index finger of the weak hand parallel to the axis of the bore of the gun.

PROPER USE OF THE SLING

When the sling is used for a non-strenuous carry, it should extend around the neck, not the strong side shoulder (opposite). Length is adjusted to allow proper underarm assault and shoulder use of weapon (center and far right).

For strenuous activities, such as rappelling, the sling should traverse the strong side shoulder. This secures the gun to the body while leaving the arms free. This is not used to assist any firing position.

SMG MAGAZINE CHANGES

Need to change magazine is realized by the operator. With the weak hand, operate the magazine release by feel, keeping your eyes downrange.

Obtain a fresh magazine.

Proceed to insert it briskly into the magazine well.

Withdraw the expended magazine from the weapon.

Place it as shown into the expended magazine bag.

Sharply rap it to ensure proper lockup.

Cycle action once to cock the weapon. This is only required on those SMGs that do not feature a bolt hold-open device. Return to action.

PIVOTS AND TURNS, 90 AND 180 DEGREES

Starting position for 90 degree left pivot.

To begin the pivot, the weak side foot is placed slightly to the rear and toward the target. Head begins to turn, fixing target.

Starting position for 90 degree right pivot.

To begin pivot, firer takes a large step across his strong foot with his weak foot. Head begins to turn toward the target.

Firer pivots on the balls of both feet toward the target, simultaneously beginning to assume the appropriate firing position.

Engage target.

Pivot on the balls of both feet. Assume appropriate firing position.

Engage target.

Starting position for 180 degree turn.

To begin turn, firer takes a sharp step to the rear, across his strong foot as shown.

MOVEMENT TOWARD SPECIFIC TARGET AREA—THE TAYLOR "SHUFFLE"

Start from the Taylor Ready position.

Body pivots on the balls of both feet. Head turns toward the target. Assume proper firing position.

Engage target.

Strong foot is carefully brought forward to a point just short of touching the weak foot. Balance is kept forward.

Weak foot is advanced about a shoulder's width toward the target. Process is repeated until the firer reaches the target area.

SITTING POSITION

Start from the Rhodesian Ready position.

KNEELING POSITION

Firer places his weak hand out and takes a step with his weak foot forward and across toward his strong side about 40 degrees.

Start in the Rhodesian Ready position.

A step is taken with the weak foot about a shoulder's width long, across an imaginary line extending from the strong foot toward the target.

The weapon is brought to bear on the target, with the torso leaning slightly forward and both elbows contacting the inside of the knees.

Firer then simply sits down, breaking his fall with his weak hand.

With a slight forward lean, engage the target. Note that the weak elbow is slightly forward of the kneecap and that the strong side elbow is down.

Firer then sits down on his strong side heel.

SGN FIRING CYCLE,
SLIDE-ACTION WEAPON

Action commences from the Taylor Ready position. Operator may either have the safety on with his trigger finger on it or have the action cracked open an inch.

The weapon is brought to bear on the target in the proper firing position.

FIRING WEARING A GAS MASK

The Taylor Underarm Assault is designed to make this easy at close range.

The shot is fired. Action is immediately cycled, ejecting the spent shell and reloading a fresh one. Note that the weapon should not be lowered.

If the problem has been solved, lower the weapon back into the Taylor Ready.

From the shoulder, the firer places the lower strong side cheek of the mask against the comb of the weapon's stock. This places his firing eye over the weapon. Target is then engaged as shown.

SGN MALFUNCTION CLEARANCE DRILLS

Failure to eject.

Firer realizes that a stoppage has occurred. He confirms the nature of the stoppage.

The shell jumps the magazine shell stop, which prevents cycle action to eject the spent shell.

Firer immediately assumes a kneeling position as shown.

He sharply "flips" the weapon to the right, allowing the expended shell to be expelled from the weapon.

He then chambers a fresh shell and resumes action.

This successfully accomplished, he completes the cycle and assumes an appropriate firing position.

He then raps the gun sharply on the deck, at the same time attempting to cycle the action to eject the spent shell in the chamber of the gun.

SMG MALFUNCTION CLEARANCE DRILLS

Trigger is pressed, and the gun fails to discharge.

Immediately rap the magazine sharply as shown.

Weapon ceases functioning, failing to eject.

Operator confirms the nature of the stoppage visually.

Draw the actuator to the rear, cocking the weapon.

Continue action.

Grasp the actuator of the weapon.

Briskly flip the weapon to the right, throwing the empty case free of the gun. Resume action.

Firer visually confirms the nature of the stoppage.

A feedway malfunction occurs.

The weapon is pointed downrange and the trigger is pressed to fire any live cartridge still chambered. Simultaneously, the magazine is either discarded or placed in the expended magazine bag, depending upon circumstances.

A fresh magazine is placed in the weapon.

He then retracts the actuator, recocking the weapon and relieving the bolt/recoil spring pressure on the stoppage.

The magazine is then withdrawn, allowing the stuck cartridges to fall free.

The actuator is drawn to the rear, recocking the weapon after the magazine is sharply rapped to ensure proper seating in the housing.

Action is resumed.

PRONE POSITION ASSUMPTION

From the forward position, pivoting on the ball of the strong side foot, turn into the strong side.

Drop to your knees, following the direction your body was pointing when you completed the pivot. Do not turn toward the target.

Thrust your weak side arm straight forward as you push your torso outward from kneeling. This breaks your fall and prevents the weapon from contacting the ground.

Roll your body back to the strong side, pulling your elbows into position as you move. Your body should be at an approximately 35 degree angle in relation to the direction of your target. Feet are spread comfortably apart with the insides of the ankles flat on the ground.

7. MODIFICATIONS AND SIGHTING ADDITIONS

There are a multitude of after-market devices and accessories available to the prospective buyer. Some common ones are optical attachments, muzzle brakes, choke devices, folding stocks and stocks made of phenolic materials, "assault slings," and spare ammunition carriers, to mention but a few. Consider your needs carefully before parting with your hard-earned cash for these items. Muzzle brakes are largely a waste of time with shotguns because they cannot exert enough gas pressure to repeal Newton's Third Law of Motion. Folding stocks are, at best, a compromise sacrificing considerable handling for an increase in storage capability in tight places. Assault slings and buttstock-mounted ammunition carriers are junk and interfere with effective weapon handling while offering nothing practical in return. If you are unable to find a buckshot load with which your gun will pattern efficiently, then the adjustable choke may well be the answer. However, you must test your gun extensively to determine its characteristics first.

While both the combat shotgun and submachine gun are best employed without optical accessories for general missions, there *are* certain situations where such devices are not only acceptable but can actually enhance efficiency. Examples of this include SWAT operations that involve entry into dimly lit interiors, thereby necessitating some sort of light attachment, or special military functions, such as raids, that might require a high volume of automatic fire directed against a small area target. For this requirement, high resolution, low magnification telescopic sights or some kind of occluded eyesight device, such as the Single Point or Armson, are in order.

Another scenario that might call for a light attachment would be any home defense situation where a shotgun is used. Normal visibility inside residential dwellings, especially at night, is poor indeed, requiring artificial illumination for successful target detection and engagement.

Remember that gadgetry is no substitute for skill and all of the gimmicks you tack onto your SGN will not cause you to shoot well. In addition, increasing the complexity of a weapon system unnecessarily also increases correspondingly the probability of a failure. For this reason it is essential that you seriously consider the issue in depth before making a final decision.

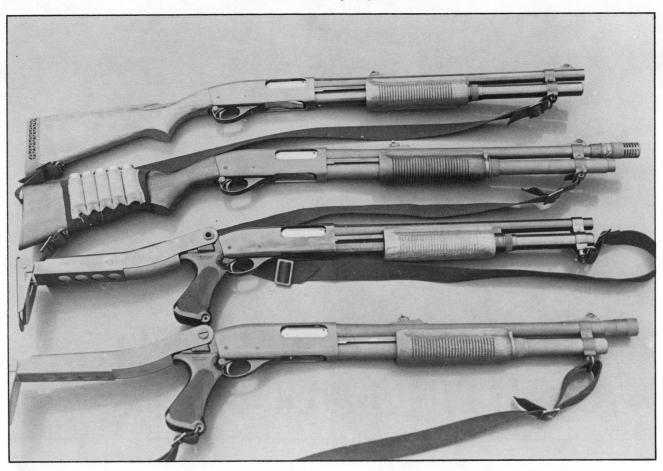

The Remington 870 with various accessories: slings, folding stocks, a buttstock-mounted ammunition carrier, adjustable chokes.

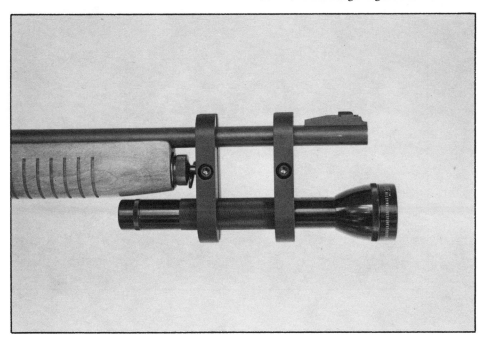

For those likely to encounter trouble in dim places, a flashlight mount is available commercially. Unfortunately, it tends to cease functioning after a few shots because the shotgun's recoil damages the contact of the flashlight bulb.

A barrel-mounted extra shell carrier may be handy, but may also be unnecessary. Think about it first.

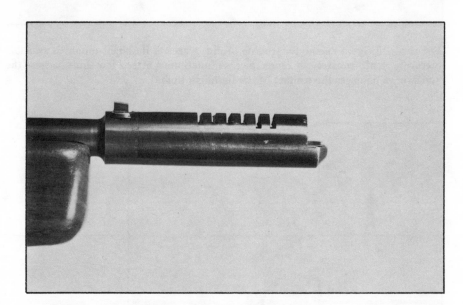

Muzzle brakes on SMGs (right and above) are largely ineffective because the relatively low gas pressures produced by SMG cartridges lack sufficient power to offset weapon vibration.

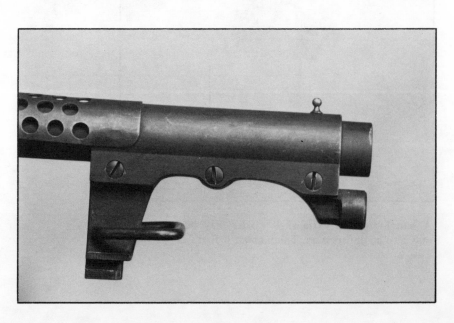

A bayonet lug and ventilated handguard assembly on your SGN might look mean, but is it really necessary, especially for $79.95!

A highly useful modification to a rifle-sighted SGN is to open up the rear notch to a configuration similar to that found on African express rifles. This allows fast sight acquisition.

There are many types of optical aids available for use with both the SGN and SMG. One of the most common is the occluded eyesight type such as the Single-Point (above) and Armson (right).

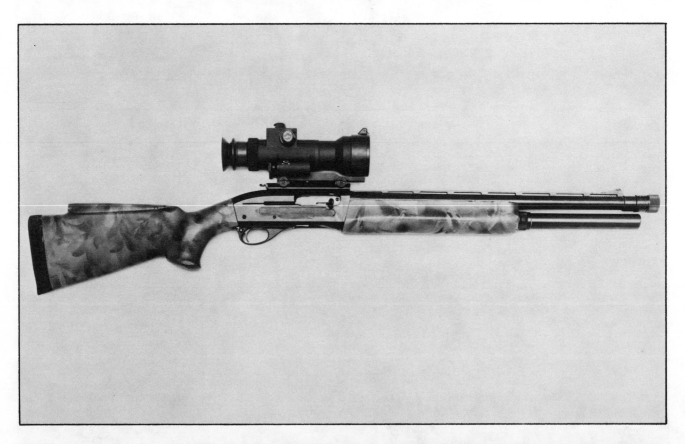

Many SMGs and SGNs can be fitted with either Infrared (far right) or Starlight (above) night vision sights for use in specialized environments. Although the passive Starlight is the more expensive and advanced of the two, both are very expensive and require considerable justification before purchase. Note that they also disturb the balance of the weapon on which they are mounted. Ponder this trade-off well before investing in these devices.

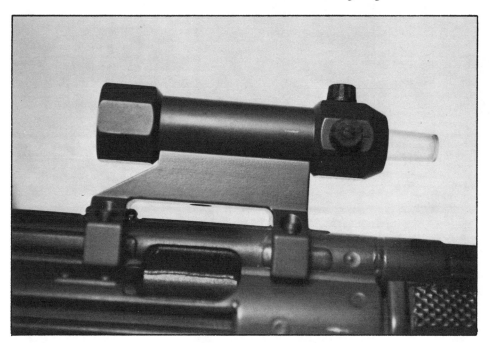

While the author finds no fault with standard configuration SGN fore ends, many prefer to increase their purchase as shown. If it helps you, do it!

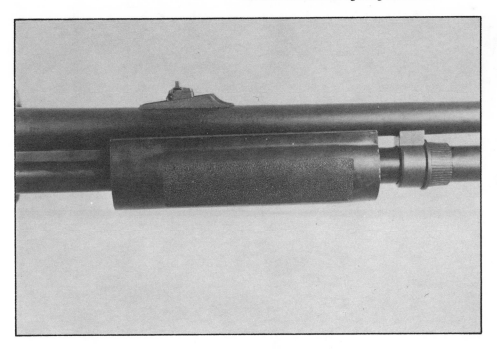

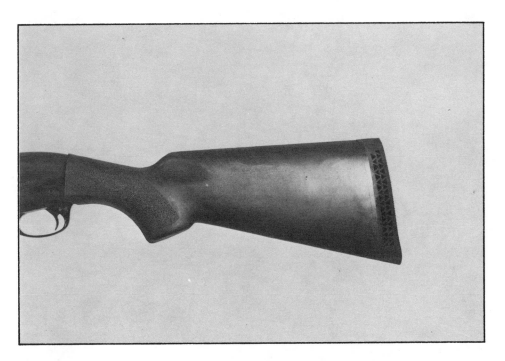

Replacement of wooden furniture with a phenolic fore end and/or buttstock is a good idea. Care must be taken, however, to ensure that all sharp edges are removed to prevent abrasion of clothing and skin.

The method of carrying spare ammunition must be determined by the purpose and the circumstances for which the weapon is to be used. This is an important accessory. Think carefully before you decide. Shown here is a tactical vest.

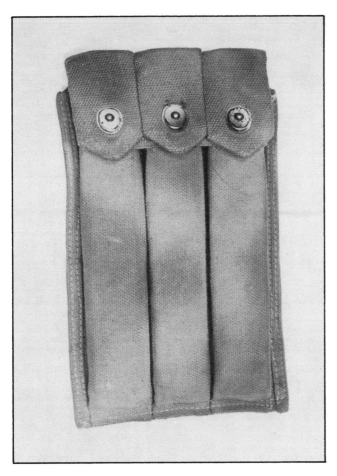

A multi-cell pouch, belt-mounted.

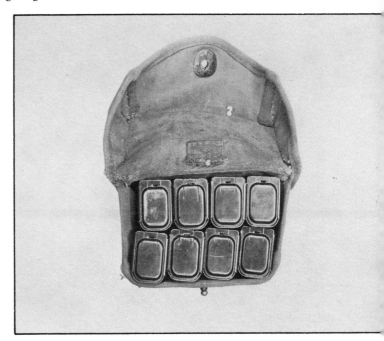

A box pouch, which can be worn either on the shoulder or waist belt.

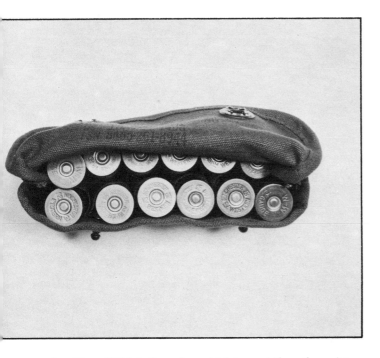

A military SGN shell carrier, which is carried on the waist.

A multi-cell pouch, shoulder-carried.

GLOSSARY

AP: Abbreviation for the term *armor-piercing*.

Armor-Piercing: A projectile designed to penetrate armor plate. Demonstrates superior penetration to standard ammunition.

AW: Abbreviation for the term *automatic weapon*.

Ball: The military designation for standard full metal jacketed projectile ammunition.

Blowback: A common type of submachine gun locking system. The action itself is held shut at the moment of firing by the inertia of the breechblock and strength of the recoil spring and is blown open upon detonation of the primer and subsequent combustion of the propellant powder housed within the cartridge case, thus cycling the action.

Bore: The interior surface of the barrel.

Breech: The rear face of the barrel. The bolt lies directly to the rear and usually acts in direct conjunction with the breech surface.

Bullet: A single projectile fired from a small arm. *A bullet is not a cartridge.*

Bullet Jacket: The material used to cover the core of the bullet. The jacket comes into contact with the rifled surfaces of the bore.

Buttstock: The rearmost stock of a shoulder weapon. The buttplate is attached to it. In turn, it is attached to the receiver of the weapon itself.

Armor-piercing ammunition. Note the hardened metal penetrator insert.

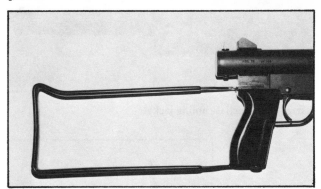

Buttstock, folding (above) and fixed (below).

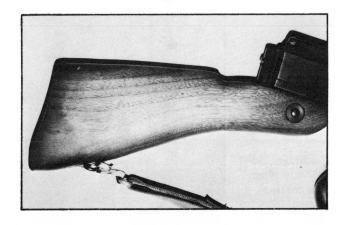

Bullets, recovered after firing.

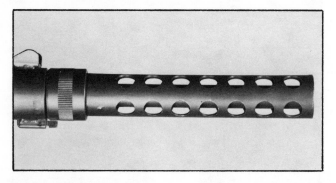

Cooling jacket.

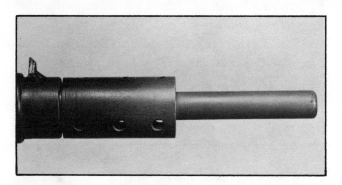

Barrel, with partial cooling jacket.

Cartridges, .30-06 and 9mm parabellum shown.

Caliber: The diameter of the bore measured to the depth of the rifling grooves. In the U.S. this is done using hundredths of an inch (.45 ACP), but in many other parts of the world, metric designations are used (9mm parabellum).

Cartridge: The complete capsule of primer, case, bullet and propellant powder. *A cartridge is not a bullet.*

Cartridge Case: The body of a cartridge. The bullet is seated inside its mouth, the propellant powder charge resides within it and the primer is seated within its base.

Centerfire: A type of cartridge in which the primer is located in the center of the case head.

Chamber: The portion of the barrel in which the cartridge resides when firing is imminent. The bullet is just to the rear of the bore and the rifling. The breech face is just to the rear of the cartridge head.

Choke: A minute reduction or constriction in the bore of a shotgun found near the muzzle which, depending upon its configuration, affects the distribution of shot pellets either positively or negatively. *Full* choke typically reduces dispersion of the shot while cylinder bore causes the opposite.

Clip: A device, normally constructed of metal, for holding a number of cartridges. Used for rapid reloading. *A clip is not a magazine.*

Cocking Piece: Also known as an *actuator;* that part of the action of a small arm that is grasped by the strong or weak hand to facilitate readying for firing.

Compensator: A device that theoretically reduces the muzzle climb of a small arm during firing. (See *muzzle brake.*)

Cooling Jacket: The portion of a firearm that encircles the barrel to aid in cooling and prevents the hand from coming into contact, avoiding burns.

Disconnector: A device that acts in conjunction with the hammer and sear. Prevents the automatic cycling of the action by interrupting the cycle of operation.

Cartridge case.

Cartridge case head, also showing primer (fired).

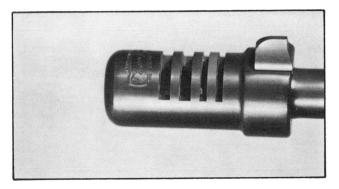

Compensator, aslo known as muzzle brake.

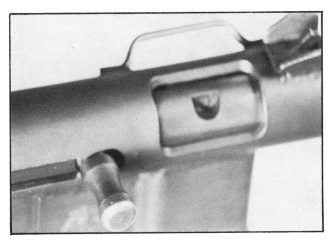

Cocking piece, also known as an actuator. Typical SMG (above) and SGN (below) types shown.

Ejector: A device which throws the spent (fired) case clear of the receiver of a firearm.

Energy: The capacity to move an object. The term used in the small arms field is "kinetic energy," and describes the striking force in foot pounds of a projectile.

Extractor: The device that withdraws the fired case from the chamber of a firearm.

Firearm: A weapon that employs internal combustion propulsion to launch a projectile or projectiles.

Firing Pin: Also known as a *striker.* A pin with a hardened, round tip which strikes the primer of a cartridge, thereby detonating it and initiating the firing cycle.

Flash Suppressor: A device attached to the muzzle of a firearm to dissipate propellant gases before they can fully exit the bore and expand into the atmosphere, thus causing flash.

Fore End: Also known as *handguard.* That portion of the weapon which is grasped by the supporting hand.

Grip Safety: A device that prevents accidental discharge of a firearm. Is usually located in the grip of the piece. It deactivates the weapon unless it is depressed by the firing hand.

Hammer: That part of a firearm that strikes the firing pin.

Incendiary: A flammable projectile designed to cause fires upon impact from the detonation of an incendiary mixture housed within the bullet.

INCEN: Abbreviation for the term *incendiary.*

Locked Breech: A type of action in which the bolt and barrel are positively locked together during the moment of firing.

Magazine: That part of a firearm in which the spare ammunition is housed. As applied to automatic weapons, the magazine fits into a housing, allowing the ammunition to be continually fed into the chamber of the weapon until it is empty. As applied to shotguns, the term usually connotes a tubular receptacle in which a number of shotshells are housed end to end. They are then fed, one at a time upon cycling of the action, into the chamber via a lifter device that brings them up into line with the barrel. *A magazine is not a clip.*

Magazine Housing: That part of the weapon in which the magazine is inserted and locked, and from which it is withdrawn when empty.

Magazine Pouch: A device for carrying spare magazines. Also known as a magazine *carrier.*

Magazine Release: That portion of the weapon which holds the magazine in the housing. It must be actuated to allow extraction of the magazine.

Muzzle: The "business end" of the barrel, from which the bullet exits as it begins flight.

Muzzle Brake: See *compensator.*

Piece: Any small arm. Derived from the term "piece of ordnance."

Power: The force exerted by a cartridge when fired from a specific firearm.

Primer: That portion of the cartridge which houses the detonating compound which ignites the propellant powder within the cartridge case.

Range: The distance to which a projectile is thrown.

Recoil: Also known as "kick." Newton's Third Law of Motion: For every action there is an equal but opposite *reaction.*

Recoil Spring: The spring that returns the bolt to its position to the rear of the breech face. During the firing cycle the spring is compressed by the rearward motion of the bolt.

Rifling: Longitudinal grooves (spirals) cut into the bore of the barrel. These impart spin to the projectile, thus stabilizing it in flight.

Rim: The flange around the cartridge case head which gives the extractor purchase.

Round: A single cartridge.

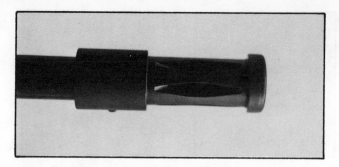

Flash suppressor.

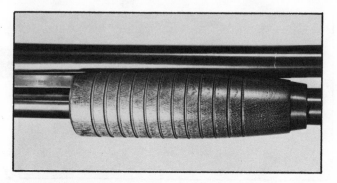

Fore end, SGN shown.

Grip safety.

Magazines. Detachable box type for SMG shown.

Sear: A part of the firearm's action, usually located in the trigger assembly, which holds the hammer in the cocked position. When the sear is forced out of engagement by the action of the trigger, the hammer falls, thus firing the weapon if it is loaded.

Selector Switch: A switch designed to allow the firer to select the mode of operation of the firearm.

SGN: Abbreviation for the term *shotgun*.

Shell: An explosive projectile hurled by a cannon or howitzer. Often used erroneously to describe a cartridge. Also used correctly to describe a shotgun cartridge.

Shot: A pellet fired from a shotgun. Also a term used to denote the discharge of a firearm.

Shotgun: A smooth bored shoulder arm that hurls a quantity of shot pellets that strike in a pattern. In special instances it can also fire a single projectile.

Shoulder Stock: A detachable stock assembly used in conjunction with a handgun, attached via a dovetail slot in the backstrap of the weapon. Also often used as a holster when not attached.

Sight: The device used to align the firer's eye with the trajectory of the bullet.

Sight Radius: The distance between the front and rear sight.

Silencer: A device used to eliminate the sound of discharge of a firearm.

Slide: That portion of a shotgun which is cycled from front to rear and back again to eject a spent shell from the weapon, recock it and chamber a fresh cartridge.

Sling: An apparatus or strap for carrying a shoulder arm. In some configurations it can also be used to assist in controlling the weapon during firing.

Small Arm: Those firearms operated by a single individual.

SMG: Abbreviation for the term *submachine gun*.

Magazine housing, Mk II STEN SMG shown.

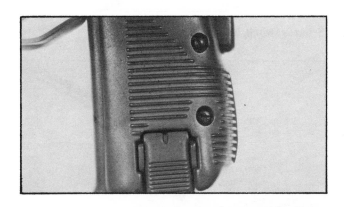

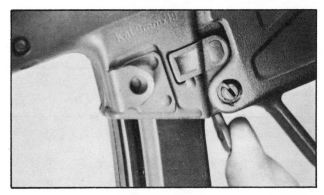

Magazine release, UZI (center), and MP-5 (above) SMGs shown.

Magazine pouch, also known as magazine carrier.

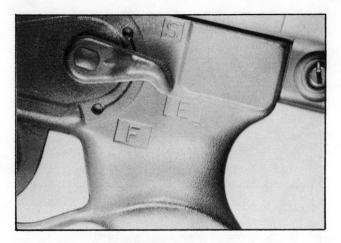

Selector switch, MP-5 SMG shown.

Sound suppressor equipped SMG (M-3A1).

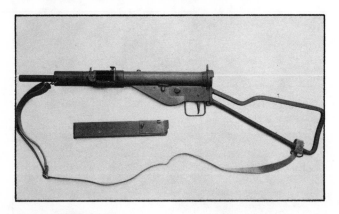

Sling on STEN SMG.

Sound Suppressor: A device to reduce significantly the sound of a discharge of a firearm. A sound suppressor is louder than a silencer.

Stock: The wood or plastic portion of the firearm intended to be held by the operator's hands.

Striker: See *firing pin*.

Strong Hand: The hand which controls the essential function of operating the weapon. If the firer is right-handed, his right hand is his strong hand.

Submachine Gun: A hand-held shoulder arm, capable of fully automatic fire, that utilizes pistol ammunition. *A submachine gun is not a machine pistol.*

Target: The object of action from a firearm and its operator.

Tracer: A projectile with a phosphorus compound located in its base ignited by propellant gases. It allows visual tracking of the projectile in flight.

TR: Abbreviation for the term *tracer*.

Trajectory: The path of a projectile through the air.

Trigger: The finger piece of the weapon which initiates the firing cycle when actuated.

Twist: The pitch of the rifling in the bore of the barrel. Normally expressed as a ratio, e.g., one turn in twelve inches, etc.

Weak Hand: The hand which is engaged primarily in supporting the weapon as opposed to actually operating it. If the firer is right-handed, his left hand is his weak hand.

Safety button, SGN type.

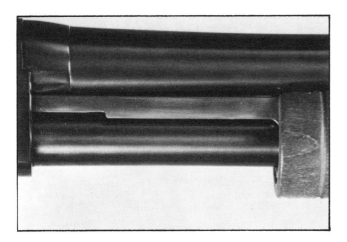

Slide.

Tracer ammunition. Note color-coded bullet tip, usually red or orange.